Food Industry: Processes and Technologies

Food Industry: Processes and Technologies

Edited by
Kaden Hunt

Larsen & Keller
www.larsen-keller.com

Food Industry: Processes and Technologies
Edited by Kaden Hunt
ISBN: 978-1-63549-123-4 (Hardback)

Larsen & Keller
Published by Larsen and Keller Education,
5 Penn Plaza,
19th Floor,
New York, NY 10001, USA

Cataloging-in-Publication Data

Food industry : processes and technologies / edited by Kaden Hunt.
p. cm.
Includes bibliographical references and index.
ISBN 978-1-63549-123-4
1. Food industry and trade. 2. Food industry and trade--Technological innovations. 3. Food. I. Hunt, Kaden.
HD9000.5 .F66 2017
338.19--dc23

The publisher's policy is to use permanent paper from mills that operate a sustainable forestry policy. Furthermore, the publisher ensures that the text paper and cover boards used have met acceptable environmental accreditation standards.

Printed and bound in the United States of America.

For more information regarding Larsen and Keller Education and its products, please visit the publisher's website www.larsen-keller.com

Table of Contents

Preface

The food industry is a global industry that aims to provide food items for consumption. The areas included in this vast industry are agriculture, foodservice operations, food processing, food manufacturing, packaging, food technology, etc. This book is compiled in such a manner, that it will provide in-depth knowledge about the theory and practice of the workings of the food industry. It sheds light on the impact of the food industry in the present day scenario. Some of the diverse topics covered in this text address the varied branches that fall under this category. This textbook, with its detailed analyses and data, will prove immensely beneficial to students involved in this area at various levels.

A foreword of all chapters of the book is provided below:

Chapter 1 - There is a rising demand for food globally due to ever growing populations. The food industry is responsible for the production, processing, storage and the sale of food products. The industry uses various techniques of mass production and processing to enable it to meet the rapidly growing food requirements. This introductory chapter introduces the reader to the food industry, food and food processing; **Chapter 2** - Specialized sectors catering to production of particular foodstuffs like dairy, confectionery, candy making, baking and snacks have been discussed in this chapter. The content discusses the methods, types, tools and cultural roles played by each sector. The chapter strategically encompasses and incorporates the major components and key concepts of food industry, providing a complete understanding; **Chapter 3** - Processed food usually contains a medley of additives in the form of food substitutes, food coloring etc. Food fortification is a method used to increase the nutrient value of food by the addition of vitamins, minerals and proteins with the aim of providing a balanced diet or overcoming dietary deficiencies. This chapter examines these food additives, their purposes, regulations related to them and their types; **Chapter 4** - Food contamination poses a serious threat as any mass produced food has a wide distribution radius. Food contamination can be caused due to faulty production techniques, improper storage or handling. This chapter deals exclusively with the topic of food contamination, foodborne illnesses and the various emerging foodborne pathogens. The chapter on food contamination offers an insightful focus, keeping in mind the complex subject matter; **Chapter 5** - Food preservation techniques stop food from spoilage and increase its shelf-life. Traditional methods of preservation like pickling of vegetables and salting or curing of meat are some of the earliest attempts at storing food for difficult times. The chapter discusses methods like pasteurization, addition of preservatives, electroporation, pascalization, food irradiation, biopreservation and food storage. The chapter strategically encompasses and incorporates the major components and key concepts of the food industry, providing a complete understanding; **Chapter 6** - Food marketing connects the producer to the consumer and involves multiple steps before the food becomes available to the buyer. This chapter explains in detail the concepts of food marketing, wholesale marketing of food and supermarkets. The reader is provided with thorough information on the types, trends, methods and representative bodies of each; **Chapter 7** - Eating out at restaurants, cafes, bistros, eateries etc. can be credited to the French Revolution. But eating a meal outside the comfort of your home was in place even before in the form of taverns and inns which served its clientele food and beverages. This chapter explores topics like foodservice, foodservice distributor and casting. The major components of food industry are discussed in this chapter; **Chapter 8** - All the food production methods and technologies can be listed under food technology. It comes as no surprise that with the leaps and bounds that science has made in the last few decades that food technology is still evolving and incorporating newer elements. This chapter discusses developments in food technology like powdered milk, freeze-drying, decaffeination and ultra-high temperature processing. A section of the content focuses on food engineering and food packaging techniques and machines. **Chapter 9** - Food security refers to the storage of food to make its supply continuous and uninterrupted. Food security is a major concern and efforts are being made to ensure every person has access to food at all times. This chapter explores the essential aspects of food security, food safety, food grading and optical sorting. It is a major part of the food industry.

At the end, I would like to thank all the people associated with this book devoting their precious time and providing their valuable contributions to this book. I would also like to express my gratitude to my fellow colleagues who encouraged me throughout the process.

Editor

1

Introduction to Food Industry

There is a rising demand for food globally due to ever growing populations. The food industry is responsible for the production, processing, storage and the sale of food products. The industry uses various techniques of mass production and processing to enable it to meet the rapidly growing food requirements. This introductory chapter introduces the reader to the food industry, food and food processing.

Food Industry

The food industry is a complex, global collective of diverse businesses that supply most of the food consumed by the world population. Only subsistence farmers, those who survive on what they grow, can be considered outside of the scope of the modern food industry.

Hens in a battery cage in Brazil, an example of intensive animal farming

The food industry includes:

- Agriculture: raising of crops and livestock, and seafood
- Manufacturing: agrichemicals, agricultural construction, farm machinery and supplies, seed, etc.
- Food processing: preparation of fresh products for market, and manufacture of prepared food products
- Marketing: promotion of generic products (e.g., milk board), new products, advertising, marketing campaigns, packaging, public relations, etc.

- Wholesale and distribution: logistics, transportation, warehousing
- Foodservice (which includes Catering)
- Grocery, farmers' markets, public markets and other retailing
- Regulation: local, regional, national, and international rules and regulations for food production and sale, including food quality, food safety, marketing/advertising, and industry lobbying activities
- Education: academic, consultancy, vocational
- Research and development: food technology
- Financial services: credit, insurance

Definitions

It is challenging to find an inclusive way to cover all aspects of food production and sale. The Food Standards Agency, a government body in India, describes it thusly:

> "...the whole food industry – from farming and food production, packaging and distribution, to retail and catering."

The Economic Research Service of the USDA uses the term *food system* to describe the same thing:

> "The U.S. food system is a complex network of farmers and the industries that link to them. Those links include makers of farm equipment and chemicals as well as firms that provide services to agribusinesses, such as providers of transportation and financial services. The system also includes the food marketing industries that link farms to consumers, and which include food and fiber processors, wholesalers, retailers, and foodservice establishments."

Agriculture and Agronomy

Agriculture is the process of producing food, feeding products, fiber and other desired products by the cultivation of certain plants and the raising of domesticated animals (livestock). The practice of agriculture is also known as "farming". Scientists, inventors, and others devoted to improving farming methods and implements are also said to be engaged in agriculture. 1 in 3 people worldwide are employed in agriculture, yet it only contributes 3% to global GDP.

A soybean field in Junin, Argentina

Agronomy is the science and technology of producing and using plants for food, fuel, fibre, and land reclamation. Agronomy encompasses work in the areas of plant genetics, plant physiology, meteorology, and soil science. Agronomy is the application of a combination of sciences. Agronomists today are involved with many issues including producing food, creating healthier food, managing environmental impact of agriculture, and extracting energy from plants.

Food Processing

Food processing includes the methods and techniques used to transform raw ingredients into food for human consumption. Food processing takes clean, harvested or slaughtered and butchered components and uses them to produce marketable food products. There are several different ways in which food can be produced.

Packaged meat in a supermarket

One off production: This method is used when customers make an order for something to be made to their own specifications, for example a wedding cake. The making of one-off products could take days depending on how intricate the design is.

Batch production: This method is used when the size of the market for a product is not clear, and where there is a range within a product line. A certain number of the same goods will be produced to make up a batch or run, for example a bakery may bake a limited number of cupcakes. This method involves estimating consumer demand.

Mass production: This method is used when there is a mass market for a large number of identical products, for example chocolate bars, ready meals and canned food. The product passes from one stage of production to another along a production line.

Just-in-time (JIT) (production): This method of production is mainly used in restaurants. All components of the product are available in-house and the customer chooses what they want in the product. It is then prepared in a kitchen, or in front of the buyer as in sandwich delicatessens, pizzerias, and sushi bars.

Wholesale and Distribution

A vast global cargo network connects the numerous parts of the industry. These include suppliers, manufacturers, warehousers, retailers and the end consumers. Wholesale markets for fresh food products have tended to decline in importance in urbanizing countries, including Latin America and some Asian countries as a result of the growth of supermarkets, which procure directly from farmers or through preferred suppliers, rather than going through markets.

A foodservice truck at a loading dock. Trucks commonly distribute food products to commercial businesses and organizations.

The constant and uninterrupted flow of product from distribution centers to store locations is a critical link in food industry operations. Distribution centers run more efficiently, throughput can be increased, costs can be lowered, and manpower better utilized if the proper steps are taken when setting up a material handling system in a warehouse.

Retail

With worldwide urbanization, food buying is increasingly removed from food production. During the 20th century, the supermarket became the defining retail element of the food industry. There, tens of thousands of products are gathered in one location, in continuous, year-round supply.

Food preparation is another area where the change in recent decades has been dramatic. Today, two food industry sectors are in apparent competition for the retail food dollar. The grocery industry sells fresh and largely raw products for consumers to use as ingredients in home cooking. The food service industry by contrast offers prepared food, either as finished products, or as partially prepared components for final "assembly". Restaurants, cafes, bakeries and mobile food trucks provide opportunities for consumers to purchase food.

Food Industry Technologies

The Passaic Agricultural Chemical Works, an agrochemical company, in Newark, New Jersey, 1876

Modern food production is defined by sophisticated technologies. These include many areas. Agricultural machinery, originally led by the tractor, has practically eliminated human labor in many areas of production. Biotechnology is driving much change, in areas as diverse as agrochemicals, plant breeding and food processing. Many other types of technology are also involved, to the point where it is hard to find an area that does not have a direct impact on the food industry. As in other fields, computer technology is also a central force, with computer networks and specialized software providing the support infrastructure to allow global movement of the myriad components involved.

Marketing

As consumers grow increasingly removed from food production, the role of product creation, advertising, and publicity become the primary vehicles for information about food. With processed food as the dominant category, marketers have almost infinite possibilities in product creation.

Labor and Education

Until the last 100 years, agriculture was labor intensive. Farming was a common occupation and millions of people were involved in food production. Farmers, largely trained from generation to generation, carried on the family business. That situation has changed dramatically today. In America in 1870, 70-80 percent of the US population was employed in agriculture. As of 2008, less than 2 percent of the population is directly employed in agriculture., and about 80% of the population lives in cities. The food industry as a complex whole requires an incredibly wide range of skills. Several hundred occupation types exist within the food industry.

Some equipment at Tartu Mill, the largest grain milling company in the Baltic states. Modern food processing factories are often highly automated and need few workers.

Food

Food is any substance consumed to provide nutritional support for the body. It is usually of plant or animal origin, and contains essential nutrients, such as fats, proteins, vitamins, or minerals. The substance is ingested by an organism and assimilated by the organism's cells to provide energy, maintain life, or stimulate growth.

Various foods

Historically, people secured food through two methods: hunting and gathering and agriculture. Today, the majority of the food energy required by the ever increasing population of the world is supplied by the food industry.

Foods from plant sources

Food safety and food security are monitored by agencies like the International Association for Food Protection, World Resources Institute, World Food Programme, Food and Agriculture Organization, and International Food Information Council. They address issues such as sustainability, biological diversity, climate change, nutritional economics, population growth, water supply, and access to food.

The right to food is a human right derived from the International Covenant on Economic, Social and Cultural Rights (ICESCR), recognizing the "right to an adequate standard of living, including adequate food", as well as the "fundamental right to be free from hunger".

Food Sources

Most food has its origin in plants. Some food is obtained directly from plants; but even animals that are used as food sources are raised by feeding them food derived from plants. Cereal grain is a staple food that provides more food energy worldwide than any other type of crop. Corn (maize), wheat, and rice – in all of their varieties – account for 87% of all grain production worldwide. Most of the grain that is produced worldwide is fed to livestock.

Some foods not from animal or plant sources include various edible fungi, especially mushrooms. Fungi and ambient bacteria are used in the preparation of fermented and pickled foods like leavened bread, alcoholic drinks, cheese, pickles, kombucha, and yogurt. Another example is blue-green algae such as Spirulina. Inorganic substances such as salt, baking soda and cream of tartar are used to preserve or chemically alter an ingredient.

Plants

Many plants and plant parts are eaten as food and around 2,000 plant species are cultivated for food. Many of these plant species have several distinct cultivars.

Seeds of plants are a good source of food for animals, including humans, because they contain the nutrients necessary for the plant's initial growth, including many healthful fats, such as Omega fats. In fact, the majority of food consumed by human beings are seed-based foods. Edible seeds include cereals (corn, wheat, rice, et cetera), legumes (beans, peas, lentils, et cetera), and nuts. Oilseeds are often pressed to produce rich oils - sunflower, flaxseed, rapeseed (including canola oil), sesame, et cetera.

Seeds are typically high in unsaturated fats and, in moderation, are considered a health food, although not all seeds are edible. Large seeds, such as those from a lemon, pose a choking hazard, while seeds from cherries and apples contain cyanide which could be poisonous only if consumed in large volumes.

Fruits are the ripened ovaries of plants, including the seeds within. Many plants and animals have coevolved such that the fruits of the former are an attractive food source to the latter, because animals that eat the fruits may excrete the seeds some distance away. Fruits, therefore, make up a significant part of the diets of most cultures. Some botanical fruits, such as tomatoes, pumpkins, and eggplants, are eaten as vegetables. (For more information, see list of fruits.)

Vegetables are a second type of plant matter that is commonly eaten as food. These include root vegetables (potatoes and carrots), bulbs (onion family), leaf vegetables (spinach and lettuce), stem vegetables (bamboo shoots and asparagus), and inflorescence vegetables (globe artichokes and broccoli and other vegetables such as cabbage or cauliflower).

Animals

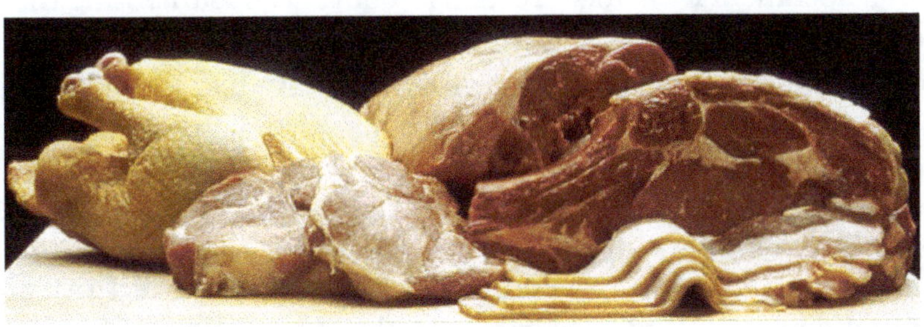

Various raw meats

Animals are used as food either directly or indirectly by the products they produce. Meat is an example of a direct product taken from an animal, which comes from muscle systems or from organs.

Food products produced by animals include milk produced by mammary glands, which in many cultures is drunk or processed into dairy products (cheese, butter, etc.). In addition, birds and other animals lay eggs, which are often eaten, and bees produce honey, a reduced nectar from flowers, which is a popular sweetener in many cultures. Some cultures consume blood, sometimes in the form of blood sausage, as a thickener for sauces, or in a cured, salted form for times of food scarcity, and others use blood in stews such as jugged hare.

Some cultures and people do not consume meat or animal food products for cultural, dietary, health, ethical, or ideological reasons. Vegetarians choose to forgo food from animal sources to varying degrees. Vegans do not consume any foods that are or contain ingredients from an animal source.

Production

Most food has always been obtained through agriculture. With increasing concern over both the methods and products of modern industrial agriculture, there has been a growing trend toward sustainable agricultural practices. This approach, partly fueled by consumer demand, encourages biodiversity, local self-reliance and organic farming methods. Major influences on food production include international organizations (e.g. the World Trade Organization and Common Agricultural Policy), national government policy (or law), and war.

A tractor and chaser bin

In popular culture, the mass production of food, specifically meats such as chicken and beef, has come under fire from various documentaries, most recently Food, Inc, documenting the mass slaughter and poor treatment of animals, often for easier revenues from large corporations. Along with a current trend towards environmentalism, people in Western culture have had an increasing trend towards the use of herbal supplements, foods for a specific group of people (such as dieters, women, or athletes), functional foods (fortified foods, such as omega-3 eggs), and a more ethnically diverse diet.

Several organisations have begun calling for a new kind of agriculture in which agroecosystems provide food but also support vital ecosystem services so that soil fertility and biodiversity are maintained rather than compromised. According to the International Water Management Institute and UNEP, well-managed agroecosystems not only provide food, fiber and animal products, they also provide services such as flood mitigation, groundwater recharge, erosion control and habitats for plants, birds fish and other animals.

Taste Perception

Animals, specifically humans, have five different types of tastes: sweet, sour, salty, bitter, and umami. As animals have evolved, the tastes that provide the most energy (sugar and fats) are the most pleasant to eat while others, such as bitter, are not enjoyable. Water, while important for survival, has no taste. Fats, on the other hand, especially saturated fats, are thicker and rich and are thus considered more enjoyable to eat.

Sweet

Generally regarded as the most pleasant taste, sweetness is almost always caused by a type of simple sugar such as glucose or fructose, or disaccharides such as sucrose, a molecule combining glucose and fructose. Complex carbohydrates are long chains and thus do not have the sweet taste. Artificial sweeteners such as sucralose are used to mimic the sugar molecule, creating the sensation of sweet, without the calories. Other types of sugar include raw sugar, which is known for its amber color, as it is unprocessed. As sugar is vital for energy and survival, the taste of sugar is pleasant.

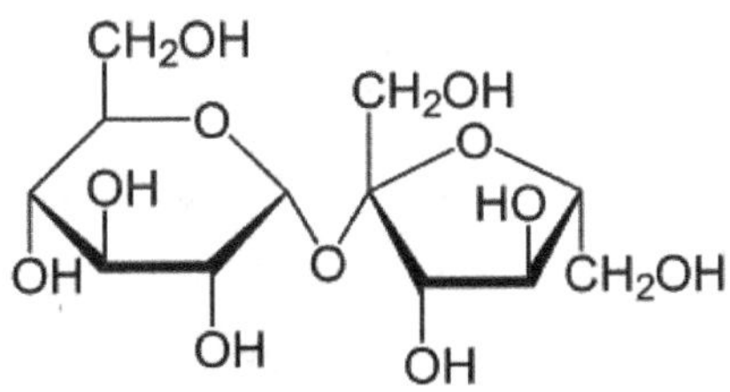

Structure of sucrose

The stevia plant contains a compound known as steviol which, when extracted, has 300 times the sweetness of sugar while having minimal impact on blood sugar.

Sour

Sourness is caused by the taste of acids, such as vinegar in alcoholic beverages. Sour foods include citrus, specifically lemons, limes, and to a lesser degree oranges. Sour is evolutionarily significant as it is a sign for a food that may have gone rancid due to bacteria. Many foods, however, are slightly acidic, and help stimulate the taste buds and enhance flavor.

Salty

Saltiness is the taste of alkali metal ions such as sodium and potassium. It is found in almost every food in low to moderate proportions to enhance flavor, although to eat pure salt is regarded as highly unpleasant. There are many different types of salt, with each having a different degree of saltiness, including sea salt, fleur de sel, kosher salt, mined salt, and grey salt. Other than enhancing flavor, its significance is that the body needs and maintains a delicate electrolyte balance, which is the kidney's function. Salt may be iodized, meaning iodine has been added to it, a necessary nutrient that promotes thyroid function. Some canned foods, notably soups or packaged broths, tend to be high in salt as a means of preserving the food longer. Historically salt has long been used as a meat preservative as salt promotes water excretion. Similarly, dried foods also promote food safety.

Salt mounds in Bolivia

Bitter

Bitterness is a sensation often considered unpleasant characterized by having a sharp, pungent taste. Unsweetened dark chocolate, caffeine, lemon rind, and some types of fruit are known to be bitter.

Umami

Umami, the Japanese word for delicious, is the least known in Western popular culture but has a long tradition in Asian cuisine. Umami is the taste of glutamates, especially monosodium glutamate (MSG). It is characterized as savory, meaty, and rich in flavor. Salmon and mushrooms are foods high in umami.

Cuisine

Many cultures have a recognizable cuisine, a specific set of cooking traditions using various spices or a combination of flavors unique to that culture, which evolves over time. Other differences include preferences (hot or cold, spicy, etc.) and practices, the study of which is known as gastronomy. Many cultures have diversified their foods by means of preparation, cooking methods, and manufacturing. This also includes a complex food trade which helps the cultures to economically survive by way of food, not just by consumption.

Some popular types of ethnic foods include Italian, French, Japanese, Chinese, American, Cajun, Thai, African, and Indian cuisine. Various cultures throughout the world study the dietary analysis of food habits. While evolutionarily speaking, as opposed to culturally, humans are omnivores, religion and social constructs such as morality, activism, or environmentalism will often affect which foods they will consume. Food is eaten and typically enjoyed through the sense of taste, the perception of flavor from eating and drinking. Certain tastes are more enjoyable than others, for evolutionary purposes.

Presentation

Aesthetically pleasing and eye-appealing food presentations can encourage people to consume foods. A common saying is that people “eat with their eyes”. Food presented in a clean and appetizing way will encourage a good flavor, even if unsatisfactory.

A French basil salmon terrine, with eye-appealing garnishes

Contrast in Texture

Texture plays a crucial role in the enjoyment of eating foods. Contrasts in textures, such as something crunchy in an otherwise smooth dish, may increase the appeal of eating it. Common examples include adding granola to yogurt, adding croutons to a salad or soup, and toasting bread to enhance its crunchiness for a smooth topping, such as jam or butter.

Contrast in Taste

Another universal phenomenon regarding food is the appeal of contrast in taste and presentation. For example, such opposite flavors as sweetness and saltiness tend to go well together, as in kettle corn and nuts.

Food Preparation

A refrigerator helps to keep foods fresh.

While many foods can be eaten raw, many also undergo some form of preparation for reasons of safety, palatability, texture, or flavor. At the simplest level this may involve washing, cutting, trimming, or adding other foods or ingredients, such as spices. It may also involve mixing, heating or cooling, pressure cooking, fermentation, or combination with other food. In a home, most food preparation takes place in a kitchen. Some preparation is done to enhance the taste or aesthetic appeal; other preparation may help to preserve the food; others may be involved in cultural identity. A meal is made up of food which is prepared to be eaten at a specific time and place.

Animal Preparation

The preparation of animal-based food usually involves slaughter, evisceration, hanging, portioning, and rendering. In developed countries, this is usually done outside the home in slaughterhouses, which are used to process animals en masse for meat production. Many countries regulate their slaughterhouses by law. For example, the United States has established the Humane Slaughter Act of 1958, which requires that an animal be stunned before killing. This act, like those in many countries, exempts slaughter in accordance to religious law, such as kosher, shechita, and dhabīĥa halal. Strict interpretations of kashrut require the animal to be fully aware when its carotid artery is cut.

On the local level, a butcher may commonly break down larger animal meat into smaller manageable cuts, and pre-wrap them for commercial sale or wrap them to order in butcher paper. In addition, fish and seafood may be fabricated into smaller cuts by a fish monger. However fish butchery may be done on board a fishing vessel and quick-frozen for preservation of quality.

Cooking

The term "cooking" encompasses a vast range of methods, tools, and combinations of ingredients to improve the flavor or digestibility of food. Cooking technique, known as culinary art, generally requires the selection, measurement, and combining of ingredients in an ordered procedure in an effort to achieve the desired result. Constraints on success include the variability of ingredients, ambient conditions, tools, and the skill of the individual cook. The diversity of cooking worldwide is a reflection of the myriad nutritional, aesthetic, agricultural, economic, cultural, and religious considerations that affect it.

Cooking with a wok in China

Cooking requires applying heat to a food which usually, though not always, chemically changes the molecules, thus changing its flavor, texture, appearance, and nutritional properties. Cooking certain proteins, such as egg whites, meats, and fish, denatures the protein, causing it to firm. There is archaeological evidence of roasted foodstuffs at Homo erectus campsites dating from 420,000 years ago. Boiling as a means of cooking requires a container, and has been practiced at least since the 10th millennium BC with the introduction of pottery.

Cooking equipment

There are many different types of equipment used for cooking.

A stainless steel frying pan

Ovens are mostly hollow devices that get very hot (up to 500 °F) and are used for baking or roasting and offer a dry-heat cooking method. Different cuisines will use different types of ovens. For example, Indian culture uses a Tandoor oven, which is a cylindrical clay oven which operates at a single high temperature. Western kitchens use variable temperature convection ovens, conventional ovens, toaster ovens, or non-radiant heat ovens like the microwave oven. Classic Italian cuisine includes the use of a brick oven containing burning wood. Ovens may be wood-fired, coal-fired, gas, electric, or oil-fired.

A traditional asado (barbecue)

Various types of cook-tops are used as well. They carry the same variations of fuel types as the ovens mentioned above. Cook-tops are used to heat vessels placed on top of the heat source, such as a sauté pan, sauce pot, frying pan, or pressure cooker. These pieces of equipment can use either a moist or dry cooking method and include methods such as steaming, simmering, boiling, and poaching for moist methods, while the dry methods include sautéing, pan frying, and deep-frying.

In addition, many cultures use grills for cooking. A grill operates with a radiant heat source from below, usually covered with a metal grid and sometimes a cover. An open pit barbecue in the American south is one example along with the American style outdoor grill fueled by wood, liquid propane, or charcoal along with soaked wood chips for smoking. A Mexican style of barbecue is called barbacoa, which involves the cooking of meats such as whole sheep over an open fire. In Argentina, an asado (Spanish for "grilled") is prepared on a grill held over an open pit or fire made upon the ground, on which a whole animal or smaller cuts are grilled.

Raw Food Preparation

Many types of fish ready to be eaten, including salmon and tuna

Certain cultures highlight animal and vegetable foods in their raw state. Salads consisting of raw vegetables or fruits are common in many cuisines. Sashimi in Japanese cuisine consists of raw sliced fish or other meat, and sushi often incorporates raw fish or seafood. Steak tartare and salmon tartare are dishes made from diced or ground raw beef or salmon, mixed with various ingredients and served with baguettes, brioche, or frites. In Italy, carpaccio is a dish of very thinly sliced raw beef, drizzled with a vinaigrette made with olive oil. The health food movement known as raw foodism promotes a mostly vegan diet of raw fruits, vegetables, and grains prepared in various ways, including juicing, food dehydration, sprouting, and other methods of preparation that do not heat the food above 118 °F (47.8 °C). An example of a raw meat dish is ceviche, a Latin American dish made with raw meat that is "cooked" from the highly acidic citric juice from lemons and limes along with other aromatics such as garlic.

Restaurants

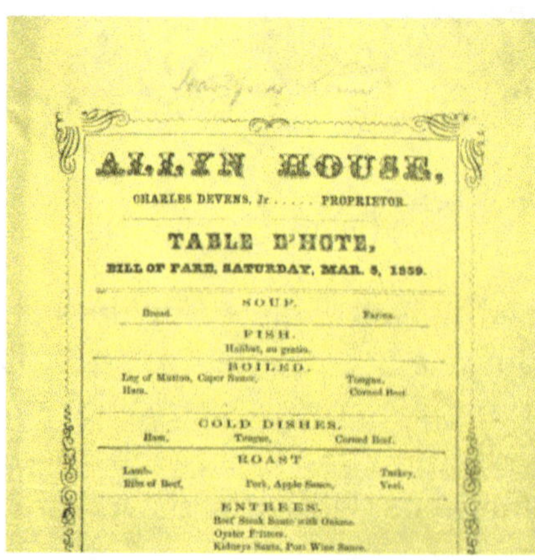

ALLYN HOUSE,

CHARLES DEVENS, Jr ..... PROPRIETOR.

TABLE D'HOTE,

BILL OF FARE, SATURDAY, MAR. 5, 1859.

SOUP.

Bread. Farina.

FISH.

Halibut, au gratin.

BOILED.

Leg of Mutton, Caper Sauce, Tongue.

Ham. Corned Beef

COLD DISHES.

Ham. Tongue. Corned Beef.

ROAST

Lamb. Turkey.

Ribs of Beef, Pork, Apple Sauce, Veal.

ENTREES.

Beef Steak Saute with Onions.

Oyster Fritters.

Kidneys Saute, Port Wine Sauce.

The Allyn House restaurant menu (March 5, 1859)

A McDonald's restaurant in Riyadh, Saudi Arabia

Restaurants employ trained chefs who prepare food, and trained waitstaff to serve the customers. The term restaurant is credited to the French from the 19th century, as it relates to the restorative nature of the bouillons that were once served in them. However, the concept pre-dates the naming of these establishments, as evidence suggests commercial food preparation may have existed during the age of the city of Pompeii, and urban sales of prepared foods may have existed in China during the Song dynasty. The coffee shops or cafés of 17th century Europe may also be considered an early version of the restaurant. In 2005, the population of the United States spent $496 billion for out-of-home dining. Expenditures by type of out-of-home dining were as follows: 40% in full-service restaurants, 37.2% in limited service restaurants (fast food), 6.6% in schools or colleges, 5.4% in bars and vending machines, 4.7% in hotels and motels, 4.0% in recreational places, and 2.2% in others, which includes military bases.

Tom's Restaurant, a restaurant in New York City

Food Manufacturing

Packaged foods are manufactured outside the home for purchase. This can be as simple as a butcher preparing meat, or as complex as a modern international food industry. Early food processing techniques were limited by available food preservation, packaging, and transportation. This mainly involved salting, curing, curdling, drying, pickling, fermenting, and smoking. Food manufacturing arose during the industrial revolution in the 19th century. This development took advantage of new mass markets and emerging technology, such as milling, preservation, packaging and labeling, and transportation. It brought the advantages of pre-prepared time-saving food to the bulk of ordinary people who did not employ domestic servants.

Packaged household food items

At the start of the 21st century, a two-tier structure has arisen, with a few international food processing giants controlling a wide range of well-known food brands. There also exists a wide array of small local or national food processing companies. Advanced technologies have also come to change food manufacture. Computer-based control systems, sophisticated processing and packaging methods, and logistics and distribution advances can enhance product quality, improve food safety, and reduce costs.

Commercial Trade

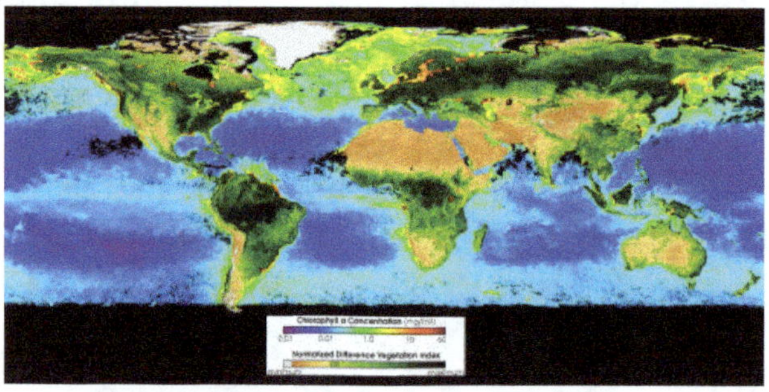

SeaWiFS image for the global biosphere

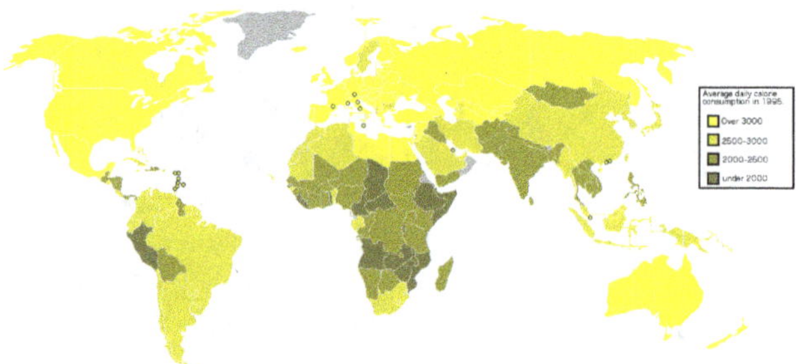

Global average daily calorie consumption in 1995

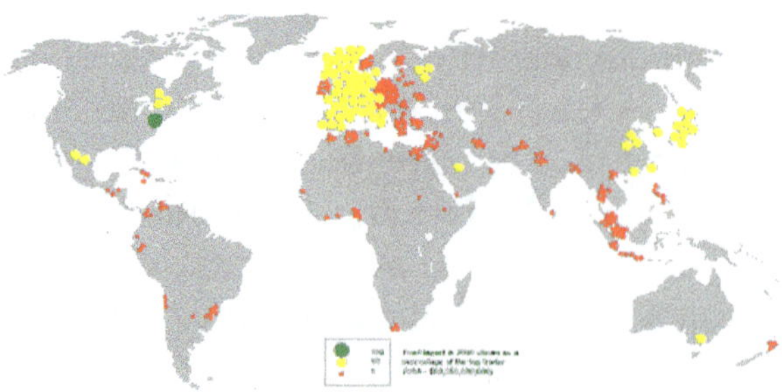

Food imports in 2005

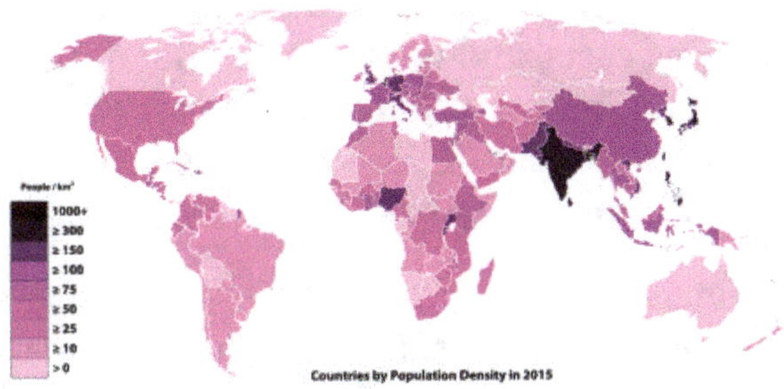

Population density of world regions

International Food Imports and Exports

The World Bank reported that the European Union was the top food importer in 2005, followed at a distance by the USA and Japan. Britain's need for food was especially well illustrated in World War II. Despite the implementation of food rationing, Britain remained dependent on food imports and the result was a long term engagement in the Battle of the Atlantic.

Food is traded and marketed on a global basis. The variety and availability of food is no longer restricted by the diversity of locally grown food or the limitations of the local growing season. Between 1961 and 1999, there was a 400% increase in worldwide food exports. Some countries are now economically dependent on food exports, which in some cases account for over 80% of all exports.

In 1994, over 100 countries became signatories to the Uruguay Round of the General Agreement on Tariffs and Trade in a dramatic increase in trade liberalization. This included an agreement to reduce subsidies paid to farmers, underpinned by the WTO enforcement of agricultural subsidy, tariffs, import quotas, and settlement of trade disputes that cannot be bilaterally resolved. Where trade barriers are raised on the disputed grounds of public health and safety, the WTO refer the dispute to the Codex Alimentarius Commission, which was founded in 1962 by the United Nations Food and Agriculture Organization and the World Health Organization. Trade liberalization has greatly affected world food trade.

Marketing and Retailing

Food marketing brings together the producer and the consumer. It is the chain of activities that brings food from "farm gate to plate". The marketing of even a single food product can be a complicated process involving many producers and companies. For example, fifty-six companies are involved in making one can of chicken noodle soup. These businesses include not only chicken and vegetable processors but also the companies that transport the ingredients and those who print labels and manufacture cans. The food marketing system is the largest direct and indirect non-government employer in the United States.

In the pre-modern era, the sale of surplus food took place once a week when farmers took their wares on market day into the local village marketplace. Here food was sold to grocers for sale in their local shops for purchase by local consumers. With the onset of industrialization and the development of the food processing industry, a wider range of food could be sold and distributed in distant locations. Typically early grocery shops would be counter-based shops, in which purchasers told the shop-keeper what they wanted, so that the shop-keeper could get it for them.

In the 20th century, supermarkets were born. Supermarkets brought with them a self service approach to shopping using shopping carts, and were able to offer quality food at lower cost through economies of scale and reduced staffing costs. In the latter part of the 20th century, this has been further revolutionized by the development of vast warehouse-sized, out-of-town supermarkets, selling a wide range of food from around the world.

Unlike food processors, food retailing is a two-tier market in which a small number of very large companies control a large proportion of supermarkets. The supermarket giants wield great purchasing power over farmers and processors, and strong influence over consumers. Nevertheless, less than 10% of consumer spending on food goes to farmers, with larger percentages going to advertising, transportation, and intermediate corporations.

Prices

It was reported on March 24, 2008, that consumers worldwide faced rising food prices. Reasons for this development include changes in the weather and dramatic changes in the global economy, including higher oil prices, lower food reserves, and growing consumer demand in China and India. In the long term, prices are expected to stabilize. Farmers will grow more grain for both fuel and food and eventually bring prices down. Already this is happening with wheat, with more crops to be planted in the United States, Canada, and Europe in 2009. However, the Food and Agriculture Organization projects that consumers still have to deal with more expensive food until at least 2018.

Some essential food products including bread, rice and pasta

It is rare for the spikes to hit all major foods in most countries at once. Food prices rose 4% in the United States in 2007, the highest increase since 1990, and are expected to climb as much again in 2008. As of December 2007, 37 countries faced food crises, and 20 had imposed some sort of food-price controls. In China, the price of pork jumped 58% in 2007. In the 1980s and 1990s, farm subsidies and support programs allowed major grain exporting countries to hold large surpluses, which could be tapped during food shortages to keep prices down. However, new trade policies have made agricultural production much more responsive to market demands, putting global food reserves at their lowest since 1983.

Food prices are rising, wealthier Asian consumers are westernizing their diets, and farmers and nations of the third world are struggling to keep up the pace. The past five years have seen rapid growth in the contribution of Asian nations to the global fluid and powdered milk manufacturing industry, which in 2008 accounted for more than 30% of production, while China alone accounts for more than 10% of both production and consumption in the global fruit and vegetable processing and preserving industry. The trend is similarly evident in industries such as soft drink and bottled water manufacturing, as well as global cocoa, chocolate, and sugar confectionery manufacturing, forecast to grow by 5.7% and 10.0% respectively during 2008 in response to soaring demand in Chinese and Southeast Asian markets.

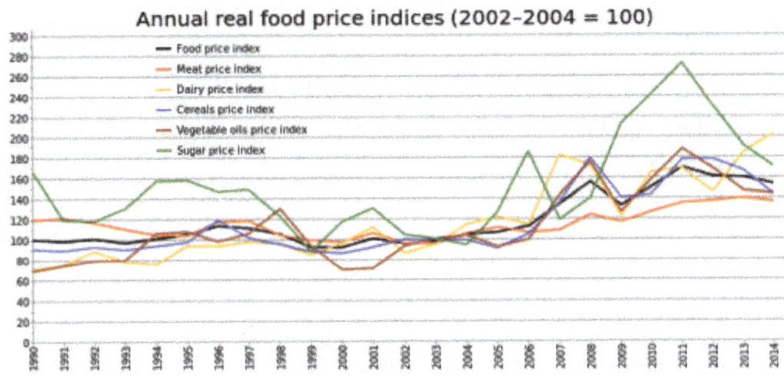

Food, meat, dairy, cereals, vegetable oil, and sugar price indices, deflated using the World Bank Manufactures Unit Value Index (MUV)

Rising food prices over recent years have been linked with social unrest around the world, including rioting in Bangladesh and Mexico, and the Arab Spring.

In 2013 Overseas Development Institute researchers showed that rice has more than doubled in price since 2000, rising by 120% in real terms. This was as a result of shifts in trade policy and restocking by major producers. More fundamental drivers of increased prices are the higher costs of fertiliser, diesel and labour. Parts of Asia see rural wages rise with potential large benefits for the 1.3 billion (2008 estimate) of Asia's poor in reducing the poverty they face. However, this negatively impacts more vulnerable groups who don't share in the economic boom, especially in Asian and African coastal cities. The researchers said the threat means social-protection policies are needed to guard against price shocks. The research proposed that in the longer run, the rises present opportunities to export for Western African farmers with high potential for rice production to replace imports with domestic production.

As Investment

Institutions such as hedge funds, pension funds and investment banks like Barclays Capital, Goldman Sachs and Morgan Stanley have been instrumental in pushing up prices in the last five years, with investment in food commodities rising from \$65bn to \$126bn (£41bn to £79bn) between 2007 and 2012, contributing to 30-year highs. This has caused price fluctuations which are not strongly related to the actual supply of food, according to the United Nations. Financial institutions now make up 61% of all investment in wheat futures. According to Olivier De Schutter, the UN special rapporteur on food, there was a rush by institutions to enter the food market following George W Bush's Commodities Futures Modernization Act of 2000. De Schutter told the Independent in March 2012: "What we are seeing now is that these financial markets have developed massively with the arrival of these new financial investors, who are purely interested in the short-term monetary gain and are not really interested in the physical thing – they never actually buy the ton of wheat or maize; they only buy a promise to buy or to sell. The result of this financialisation of the commodities market is that the prices of the products respond increasingly to a purely speculative logic. This explains why in very short periods of time we see prices spiking or bubbles exploding, because prices are less and less determined by the real match between supply and demand." In 2011, 450 economists from around the world called on the G20 to regulate the commodities market more.

Some experts have said that speculation has merely aggravated other factors, such as climate change, competition with bio-fuels and overall rising demand. However, some such as Jayati Ghosh, professor of economics at Jawaharlal Nehru University in New Delhi, have pointed out that prices have increased irrespective of supply and demand issues: Ghosh points to world wheat prices, which doubled in the period from June to December 2010, despite there being no fall in global supply.

Famine and Hunger

Food deprivation leads to malnutrition and ultimately starvation. This is often connected with famine, which involves the absence of food in entire communities. This can have a devastating and widespread effect on human health and mortality. Rationing is sometimes used to distribute food in times of shortage, most notably during times of war.

Starvation is a significant international problem. Approximately 815 million people are undernourished, and over 16,000 children die per day from hunger-related causes. Food deprivation is regarded as a deficit need in Maslow's hierarchy of needs and is measured using famine scales.

Food Aid

Food aid can benefit people suffering from a shortage of food. It can be used to improve peoples' lives in the short term, so that a society can increase its standard of living to the point that food aid is no longer required. Conversely, badly managed food aid can create problems by disrupting local markets, depressing crop prices, and discouraging food production. Sometimes a cycle of food aid dependence can develop. Its provision, or threatened withdrawal, is sometimes used as a political tool to influence the policies of the destination country, a strategy known as food politics. Sometimes, food aid provisions will require certain types of food be purchased from certain sellers, and food aid can be misused to enhance the markets of donor countries. International efforts to distribute food to the neediest countries are often coordinated by the World Food Programme.

Safety

Foodborne illness, commonly called "food poisoning", is caused by bacteria, toxins, viruses, parasites, and prions. Roughly 7 million people die of food poisoning each year, with about 10 times as many suffering from a non-fatal version. The two most common factors leading to cases of bacterial foodborne illness are cross-contamination of ready-to-eat food from other uncooked foods and improper temperature control. Less commonly, acute adverse reactions can also occur if chemical contamination of food occurs, for example from improper storage, or use of non-food grade soaps and disinfectants. Food can also be adulterated by a very wide range of articles (known as "foreign bodies") during farming, manufacture, cooking, packaging, distribution, or sale. These foreign bodies can include pests or their droppings, hairs, cigarette butts, wood chips, and all manner of other contaminants. It is possible for certain types of food to become contaminated if stored or presented in an unsafe container, such as a ceramic pot with lead-based glaze.

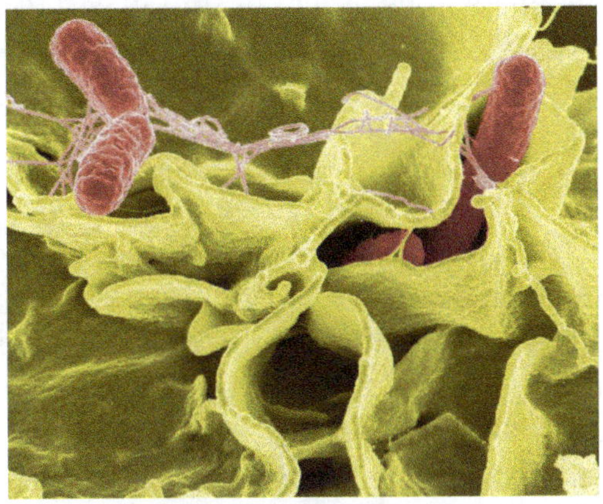

Salmonella bacteria is a common cause of foodborne illness, particularly in undercooked chicken and chicken eggs.

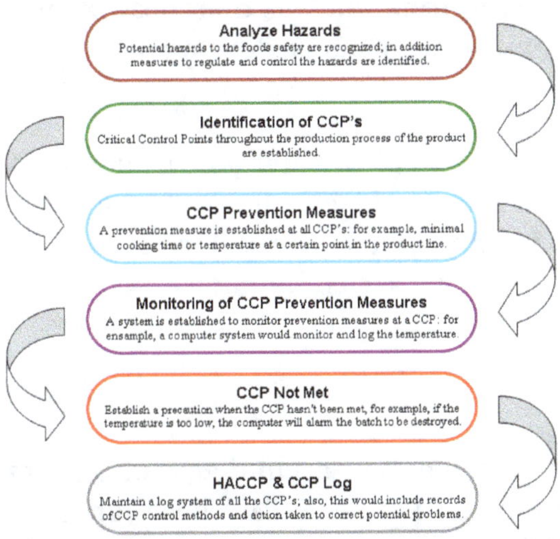

Hazard Analysis and Critical Control Points (HACCP) Flowchart

Food poisoning has been recognized as a disease since as early as Hippocrates. The sale of rancid, contaminated, or adulterated food was commonplace until the introduction of hygiene, refrigeration, and vermin controls in the 19th century. Discovery of techniques for killing bacteria using heat, and other microbiological studies by scientists such as Louis Pasteur, contributed to the modern sanitation standards that are ubiquitous in developed nations today. This was further underpinned by the work of Justus von Liebig, which led to the development of modern food storage and food preservation methods. In more recent years, a greater understanding of the causes of food-borne illnesses has led to the development of more systematic approaches such as the Hazard Analysis and Critical Control Points (HACCP), which can identify and eliminate many risks.

Recommended measures for ensuring food safety include maintaining a clean preparation area with foods of different types kept separate, ensuring an adequate cooking temperature, and refrigerating foods promptly after cooking.

Foods that spoil easily, such as meats, dairy, and seafood, must be prepared a certain way to avoid contaminating the people for whom they are prepared. As such, the rule of thumb is that cold foods (such as dairy products) should be kept cold and hot foods (such as soup) should be kept hot until storage. Cold meats, such as chicken, that are to be cooked should not be placed at room temperature for thawing, at the risk of dangerous bacterial growth, such as Salmonella or E. coli.

Allergies

Some people have allergies or sensitivities to foods which are not problematic to most people. This occurs when a person's immune system mistakes a certain food protein for a harmful foreign agent and attacks it. About 2% of adults and 8% of children have a food allergy. The amount of the food substance required to provoke a reaction in a particularly susceptible individual can be quite small. In some instances, traces of food in the air, too minute to be perceived through smell, have been known to provoke lethal reactions in extremely sensitive individuals. Common food allergens are gluten, corn, shellfish (mollusks), peanuts, and soy. Allergens frequently produce symptoms such as diarrhea, rashes, bloating, vomiting, and regurgitation. The digestive complaints usually develop within half an hour of ingesting the allergen.

Rarely, food allergies can lead to a medical emergency, such as anaphylactic shock, hypotension (low blood pressure), and loss of consciousness. An allergen associated with this type of reaction is peanut, although latex products can induce similar reactions. Initial treatment is with epinephrine (adrenaline), often carried by known patients in the form of an Epi-pen or Twinject.

Other Health Issues

Human diet was estimated to cause perhaps around 35% of cancers in a human epidemiological analysis by Richard Doll and Richard Peto in 1981. These cancer may be caused by carcinogens that are present in food naturally or as contaminants. Food contaminated with fungal growth may contain mycotoxins such as aflatoxins which may be found in contaminated corn and peanuts. Other carcinogens identified in food include heterocyclic amines generated in meat when cooked at high temperature, polyaromatic hydrocarbons in charred meat and smoked fish, and nitrosamines generated from nitrites used as food preservatives in cured meat such as bacon.

Anticarcinogens that may help prevent cancer can also be found in many food especially fruits and vegetable. Antioxidants are important groups of compounds that may help remove potentially harmful chemicals. It is however often difficult to identify the specific components in diet that serve to increase or decrease cancer risk since many food, such as beef steak and broccoli, contain low concentrations of both carcinogens and anticarcinogens.

Diet

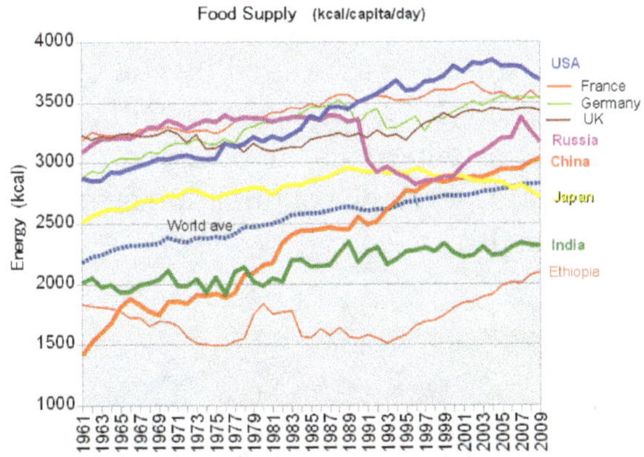

Changes of food supply (by energy) Other area (Yr 2010) * Africa, sub-Sahara - 2170 kcal/capita/day * N.E. and N. Africa - 3120 kcal/capita/day * South Asia - 2450 kcal/capita/day * East Asia - 3040 kcal/capita/day * Latin America / Caribbean - 2950 kcal/capita/day * Developed countries - 3470 kcal/capita/day

Cultural and Religious Diets

Dietary habits are the habitual decisions a person or culture makes when choosing what foods to eat. Many cultures hold some food preferences and some food taboos. Dietary choices can also define cultures and play a role in religion. For example, only kosher foods are permitted by Judaism, halal foods by Islam, and in Hinduism beef is restricted. In addition, the dietary choices of different countries or regions have different characteristics. This is highly related to a culture's cuisine.

Diet Deficiencies

Dietary habits play a significant role in the health and mortality of all humans. Imbalances between the consumed fuels and expended energy results in either starvation or excessive reserves of adipose tissue, known as body fat. Poor intake of various vitamins and minerals can lead to diseases that can have far-reaching effects on health. For instance, 30% of the world's population either has, or is at risk for developing, iodine deficiency. It is estimated that at least 3 million children are blind due to vitamin A deficiency. Vitamin C deficiency results in scurvy. Calcium, Vitamin D, and phosphorus are inter-related; the consumption of each may affect the absorption of the others. Kwashiorkor and marasmus are childhood disorders caused by lack of dietary protein.

Moral, Ethical, and Health-Conscious Diets

Many individuals limit what foods they eat for reasons of morality, or other habit. For instance, vegetarians choose to forgo food from animal sources to varying degrees. Others choose a healthier diet, avoiding sugars or animal fats and increasing consumption of dietary fiber and antioxidants. Obesity, a serious problem in the western world, leads to higher chances of developing heart disease, diabetes, and many other diseases. More recently, dietary habits have been influenced by the concerns that some people have about possible impacts on health or the environment from genetically modified food. Further concerns about the impact of industrial farming (grains) on animal welfare, human health, and the environment are also having an effect on contemporary human dietary habits. This has led to the emergence of a movement with a preference for organic and local food.

Nutrition and Dietary Problems

Between the extremes of optimal health and death from starvation or malnutrition, there is an array of disease states that can be caused or alleviated by changes in diet. Deficiencies, excesses, and imbalances in diet can produce negative impacts on health, which may lead to various health problems such as scurvy, obesity, or osteoporosis, diabetes, cardiovascular diseases as well as psychological and behavioral problems. The science of nutrition attempts to understand how and why specific dietary aspects influence health.

MyPlate replaced MyPyramid as the USDA nutrition guide.

Nutrients in food are grouped into several categories. Macronutrients are fat, protein, and carbohydrates. Micronutrients are the minerals and vitamins. Additionally, food contains water and dietary fiber.

As previously discussed, the body is designed by natural selection to enjoy sweet and fattening foods for evolutionary diets, ideal for hunters and gatherers. Thus, sweet and fattening foods in nature are typically rare and are very pleasurable to eat. In modern times, with advanced technology, enjoyable foods are easily available to consumers. Unfortunately, this promotes obesity in adults and children alike.

Legal Definition

Some countries list a legal definition of food, often referring them with the word *foodstuff*. These countries list food as any item that is to be processed, partially processed, or unprocessed for consumption. The listing of items included as food include any substance intended to be, or reasonably expected to be, ingested by humans. In addition to these foodstuffs, drink, chewing gum, water, or other items processed into said food items are part of the legal definition of food. Items not included in the legal definition of food include animal feed, live animals (unless being prepared for sale in a market), plants prior to harvesting, medicinal products, cosmetics, tobacco and tobacco products, narcotic or psychotropic substances, and residues and contaminants.

Food Processing

Food processing is the transformation of raw ingredients, by physical or chemical means into food, or of food into other forms. Food processing combines raw food ingredients to produce marketable food products that can be easily prepared and served by the consumer. Food processing typically involves activities such as mincing and macerating, liquefaction, emulsification, and cooking (such as boiling, broiling, frying, or grilling); pickling, pasteurization, and many other kinds of preservation; and canning or other packaging. (Primary-processing such as dicing or slicing, freezing or drying when leading to secondary products are also included.)

Industrial cheese production

Michael Foods egg-processing plant in Wakefield, Nebraska

History

Food processing dates back to the prehistoric ages when crude processing incorporated fermenting, sun drying, preserving with salt, and various types of cooking (such as roasting, smoking, steaming, and oven baking), Such basic food processing involved chemical enzymatic changes to the basic structure of food in its natural form, as well served to build a barrier against surface microbial activity that caused rapid decay. Salt-preservation was especially common for foods that constituted warrior and sailors' diets until the introduction of canning methods. Evidence for the existence of these methods can be found in the writings of the ancient Greek, Chaldean, Egyptian and Roman civilizations as well as archaeological evidence from Europe, North and South America and Asia. These tried and tested processing techniques remained essentially the same until the advent of the industrial revolution. Examples of ready-meals also date back to before the preindustrial revolution, and include dishes such as Cornish pasty and Haggis. Both during ancient times and today in modern society these are considered processed foods.

Modern food processing technology developed in the 19th and 20th centuries was developed in a large part to serve military needs. In 1809 Nicolas Appert invented a hermetic bottling technique that would preserve food for French troops which ultimately contributed to the development of tinning, and subsequently canning by Peter Durand in 1810. Although initially expensive and somewhat hazardous due to the lead used in cans, canned goods would later become a staple around the world. Pasteurization, discovered by Louis Pasteur in 1864, improved the quality of preserved foods and introduced the wine, beer, and milk preservation.

Grain silos in Ardrossan, Scotland

In the 20th century, World War II, the space race and the rising consumer society in developed countries (including the United States) contributed to the growth of food processing with such advances as spray drying, juice concentrates, freeze drying and the introduction of artificial sweeteners, colouring agents, and such preservatives as sodium benzoate. In the late 20th century, products such as dried instant soups, reconstituted fruits and juices, and self cooking meals such as MRE food ration were developed.

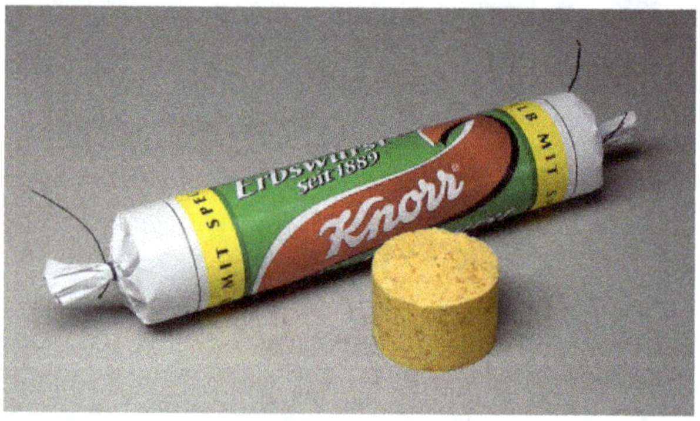

A form of pre-made split-pea soup that has become traditional

In western Europe and North America, the second half of the 20th century witnessed a rise in the pursuit of convenience. Food processing companies marketed their products especially towards middle-class working wives and mothers. Frozen foods (often credited to Clarence Birdseye) found their success in sales of juice concentrates and "TV dinners". Processors utilised the perceived value of time to appeal to the postwar population, and this same appeal contributes to the success of convenience foods today.

Benefits and Drawbacks

Benefits

Benefits of food processing include toxin removal, preservation, easing marketing and distribution tasks, and increasing food consistency. In addition, it increases yearly availability of many foods, enables transportation of delicate perishable foods across long distances and makes many kinds of foods safe to eat by de-activating spoilage and pathogenic micro-organisms. Modern supermarkets would not exist without modern food processing techniques, and long voyages would not be possible.

Processed foods are usually less susceptible to early spoilage than fresh foods and are better suited for long distance transportation from the source to the consumer. When they were first introduced, some processed foods helped to alleviate food shortages and improved the overall nutrition of populations as it made many new foods available to the masses.

Processed seafood - fish, squid, prawn balls and simulated crab sticks (surimi)

Processing can also reduce the incidence of food borne disease. Fresh materials, such as fresh produce and raw meats, are more likely to harbour pathogenic micro-organisms (e.g. Salmonella) capable of causing serious illnesses.

The extremely varied modern diet is only truly possible on a wide scale because of food processing. Transportation of more exotic foods, as well as the elimination of much hard labour gives the modern eater easy access to a wide variety of food unimaginable to their ancestors.

The act of processing can often improve the taste of food significantly.

Libby's brand 'Potted Meat Food Product'

Mass production of food is much cheaper overall than individual production of meals from raw ingredients. Therefore, a large profit potential exists for the manufacturers and suppliers of processed food products. Individuals may see a benefit in convenience, but rarely see any direct financial cost benefit in using processed food as compared to home preparation.

Processed food freed people from the large amount of time involved in preparing and cooking "natural" unprocessed foods. The increase in free time allows people much more choice in life style than previously allowed. In many families the adults are working away from home and therefore there is little time for the preparation of food based on fresh ingredients. The food industry offers products that fulfill many different needs: e.g. fully prepared ready meals that can be heated up in the microwave oven within a few minutes.

Modern food processing also improves the quality of life for people with allergies, diabetics, and other people who cannot consume some common food elements. Food processing can also add extra nutrients such as vitamins.

Drawbacks

Any processing of food can affect its nutritional density. The amount of nutrients lost depends on the food and processing method. For example, heat destroys vitamin C. Therefore, canned fruits possess less vitamin C than their fresh alternatives. The USDA conducted a study in 2004, creating a nutrient retention table for several foods.

New research highlighting the importance to human health of a rich microbial environment in the intestine indicates that abundant food processing (not fermentation of foods) endangers that environment.

Meat packages in a Roman supermarket

Using food additives represents another safety concern. The health risks of any given additive vary greatly from person to person; for example using sugar as an additive endangers diabetics. In the European Union, only European Food Safety Authority (EFSA) approved food additives (e.g., sweeteners, preservatives, stabilizers) are permitted at specified levels for use in food products. Approved additives receive an E number (E for Europe), simplifying communication about food additives included in the ingredients' list for all the different languages spoken in the EU. As effects of chemical additives are learnt, changes to laws and regulatory practices are made to make such processed foods more safe.

Food processing is typically a mechanical process that utilizes large mixing, grinding, chopping and emulsifying equipment in the production process. These processes inherently introduce a number of contamination risks. As a mixing bowl or grinder is used over time the food contact parts will tend to fail and fracture. This type of failure will introduce into the product stream small to large metal contaminants. Further processing of these metal fragments will result in downstream equipment failure and the risk of ingestion by the consumer. Food manufacturers utilize industrial metal detectors to detect and reject automatically any metal fragment. Large food processors will utilize many metal detectors within the processing stream to reduce both damage to processing machinery as well as risk to consumer health.

Typical Maximum Nutrient Losses due to cooking					
Vitamin & Minerals	**Freeze**	**Dry**	**Cook**	**Cook+Drain**	**Reheat**
Vitamin A	5%	50%	25%	35%	10%
Vit A- Retinol Activity Equivalent	5%	50%	25%	35%	10%
Vit A- Alpha Carotene	5%	50%	25%	35%	10%
Vit A- Beta Carotene	5%	50%	25%	35%	10%
Vit A- Beta Cryptoxanthin	5%	50%	25%	35%	10%
Vit A- Lycopene	5%	50%	25%	35%	10%
Vit A- Lutein+Zeaxanthin	5%	50%	25%	35%	10%
Vitamin C	30%	80%	50%	75%	50%
Thiamin	5%	30%	55%	70%	40%
Riboflavin	0%	10%	25%	45%	5%
Niacin	0%	10%	40%	55%	5%
Vitamin B6	0%	10%	50%	65%	45%
Folate	5%	50%	70%	75%	30%
Food Folate	5%	50%	70%	75%	30%
Folic Acid	5%	50%	70%	75%	30%
Vitamin B12	0%	0%	45%	50%	45%
Calcium	5%	0%	20%	25%	0%
Iron	0%	0%	35%	40%	0%
Magnesium	0%	0%	25%	40%	0%
Phosphorus	0%	0%	25%	35%	0%
Potassium	10%	0%	30%	70%	0%
Sodium	0%	0%	25%	55%	0%
Zinc	0%	0%	25%	25%	0%
Copper	10%	0%	40%	45%	0%

Performance Parameters for Food Processing

Factory automation - robotics palettizing bread

When designing processes for the food industry the following performance parameters may be taken into account:

- Hygiene, e.g. measured by number of micro-organisms per mL of finished product
- Energy efficiency measured e.g. by "ton of steam per ton of sugar produced"
- Minimization of waste, measured e.g. by "percentage of peeling loss during the peeling of potatoes"
- Labour used, measured e.g. by "number of working hours per ton of finished product"
- Minimization of cleaning stops measured e.g. by "number of hours between cleaning stops"

De-agglomerating Batter Mixes in Food Processing

Problems often occur during preparation of batter mixes because flour and other powdered ingredients tend to form lumps or agglomerates as they are being mixed during production. A conventional mixer/ agitator cannot break down these agglomerates, resulting in a lumpy batter. If lumpy batter is used to enrobe products, it causes an unsatisfactory appearance with misshapen or oversize products that do not fit properly into packaging. This can force production to a standstill. Furthermore, batter mix is generally recirculated from an enrobing system back to a holding vessel; lumps then have a tendency to build up, reducing the flow of material and raising potential sanitation issues.

Using a high shear in-line mixer in place of a conventional agitator or mixer can quickly solve problems of agglomeration with dry ingredients. A single pass through a self-pumping, in-line mixer adds high shear to batter, which de-agglomerates the mix, resulting in a homogeneous, smooth batter. With a consistent, smooth batter, finished product appearance is improved; the effectiveness and hygiene of the recirculation system is increased; and a better yield of raw materials is achieved. By increasing overall product quality, the amount of raw materials needed is decreased, thereby lowering manufacturing costs and increasing shelf life. Increased shelf life is achieved by creating and maintaining an emulsion, often by adding a food stabilizer.

High shear in-line mixers process food to be made faster and cheaper while increasing consistency of the finished food. Powder and liquid mixing systems are capable of rapidly incorporating large quantities of powders at high concentrations – agglomerate free and fully hydrated. Advances in technology have made processing equipment easy to clean, leading to a much safer processed food.

Trends in Modern Food Processing

Women working in a cannery

Dried bananas packaged in Ban Bang Krathum, Bang Krathum, Phitsanulok, Thailand

Health

- Reduction of fat content in final product by using baking instead of deep-frying in the production of potato chips, another processed food.
- Maintaining the natural taste of the product by using less artificial sweetener than was used before.

Hygiene

The rigorous application of industry and government endorsed standards to minimise possible risk and hazards. The international standard adopted is HACCP.

Efficiency

- Rising energy costs lead to increasing usage of energy-saving technologies, e.g. frequency converters on electrical drives, heat insulation of factory buildings and heated vessels, energy recovery systems, keeping a single fish frozen all the way from China to Switzerland.
- Factory automation systems (often Distributed control systems) reduce personnel costs and may lead to more stable production results.

References

- Aguilera, Jose Miguel and David W. Stanley. Microstructural Principles of Food Processing and Engineering. Springer, 1999. ISBN 0-8342-1256-0.
- Campbell, Bernard Grant. Human Evolution: An Introduction to Man's Adaptations. Aldine Transaction: 1998. ISBN 0-202-02042-8.
- Humphery, Kim. Shelf Life: Supermarkets and the Changing Cultures of Consumption. Cambridge University Press, 1998. ISBN 0-521-62630-7.
- Magdoff, Fred; Foster, John Bellamy; and Buttel, Frederick H. Hungry for Profit: The Agribusiness Threat to Farmers, Food, and the Environment. September 2000. ISBN 1-58367-016-5.
- McGee, Harold. On Food and Cooking: The Science and Lore of the Kitchen. New York: Simon & Schuster, 2004. ISBN 0-684-80001-2.
- Messer, Ellen; Derose, Laurie Fields and Sara Millman. Who's Hungry? and How Do We Know?: Food Shortage, Poverty, and Deprivation. United Nations University Press, 1998. ISBN 92-808-0985-7.

- Parekh, Sarad R. The Gmo Handbook: Genetically Modified Animals, Microbes, and Plants in Biotechnology. Humana Press,2004. ISBN 1-58829-307-6.
- Schor, Juliet; Taylor, Betsy (editors). Sustainable Planet: Roadmaps for the Twenty-First Century. Beacon Press, 2003. ISBN 0-8070-0455-3.
- Van den Bossche, Peter. The Law and Policy of the bosanac Trade Organization: Text, Cases and Materials. UK: Cambridge University Press, 2005. ISBN 0-521-82290-4.
- "Solutions for Your Toughest Mixing Applications in Food De-agglomerating" (PDF). Silverson Machines Application Report. Retrieved 2012-02-29.

2

Sectors of Food Industry

Specialized sectors catering to production of particular foodstuffs like dairy, confectionery, candy making, baking and snacks have been discussed in this chapter. The content discusses the methods, types, tools and cultural roles played by each sector. The chapter strategically encompasses and incorporates the major components and key concepts of food industry, providing a complete understanding.

Dairy

A dairy is a business enterprise established for the harvesting or processing (or both) of animal milk – mostly from cows or goats, but also from buffaloes, sheep, horses, or camels – for human consumption. A dairy is typically located on a dedicated dairy farm or in a section of a multi-purpose farm (mixed farm) that is concerned with the harvesting of milk.

Old mountain pasture dairy in Schröcken, Vorarlberg, Austria, in the Bregenz Forest

Terminology differs between countries. For example, in the United States, the entire dairy farm is commonly called a "dairy." The building or farm area where milk is harvested from the cow is often called a "milking parlor" or "parlor." The farm area where milk is stored in bulk tanks is known as the farm's "milk house." Milk is then hauled (usually by truck) to a "dairy plant," also referred to as a "dairy", where raw milk is further processed and prepared for commercial sale of dairy products. In New Zealand, farm areas for milk harvesting are also called "milking parlours", and are historically known as "milking sheds." As in the United States, sometimes milking sheds are referred to by their type, such as "herring bone shed" or "pit parlour". Parlour design has evolved from simple barns or sheds to large rotary structures in which the workflow (throughput of cows) is very efficiently handled. In some countries, especially those with small numbers of animals being milked, the farm may perform the functions of a dairy plant, processing their own milk into salable dairy products, such as butter, cheese, or yogurt. This on-site processing is a traditional method of producing specialist milk products, common in Europe.

In the United States a *dairy* can also be a place that processes, distributes and sells dairy products, or a room, building or establishment where milk is stored and processed into milk products, such as butter or cheese. In New Zealand English the singular use of the word *dairy* almost exclusively refers to a corner shop, or superette. This usage is historical as such shops were a common place for the public to buy milk products.

As an attributive, the word *dairy* refers to milk-based products, derivatives and processes, and the animals and workers involved in their production: for example dairy cattle, dairy goat. A dairy farm produces milk and a dairy factory processes it into a variety of dairy products. These establishments constitute the global dairy industry, a component of the food industry.

History

Milk producing animals have been domesticated for thousands of years. Initially, they were part of the subsistence farming that nomads engaged in. As the community moved about the country, their animals accompanied them. Protecting and feeding the animals were a big part of the symbiotic relationship between the animals and the herders.

In the more recent past, people in agricultural societies owned dairy animals that they milked for domestic and local (village) consumption, a typical example of a cottage industry. The animals might serve multiple purposes (for example, as a draught animal for pulling a plough as a youngster, and at the end of its useful life as meat). In this case the animals were normally milked by hand and the herd size was quite small, so that all of the animals could be milked in less than an hour—about 10 per milker. These tasks were performed by a *dairymaid* (*dairywoman*) or *dairyman*. The word *dairy* harkens back to Middle English *dayerie, deyerie,* from *deye* (female servant or dairymaid) and further back to Old English *dæge* (kneader of bread).

With industrialisation and urbanisation, the supply of milk became a commercial industry, with specialised breeds of cattle being developed for dairy, as distinct from beef or draught animals. Initially, more people were employed as milkers, but it soon turned to mechanisation with machines designed to do the milking.

Historically, the milking and the processing took place close together in space and time: on a dairy farm. People milked the animals by hand; on farms where only small numbers are kept, hand-milking may still be practiced. Hand-milking is accomplished by grasping the teats (often pronounced *tit* or *tits*) in the hand and expressing milk either by squeezing the fingers progressively, from the udder end to the tip, or by squeezing the teat between thumb and index finger, then moving the hand downward from udder towards the end of the teat. The action of the hand or fingers is designed to close off the milk duct at the udder (upper) end and, by the movement of the fingers, close the duct progressively to the tip to express the trapped milk. Each half or quarter of the udder is emptied one milk-duct capacity at a time.

The *stripping* action is repeated, using both hands for speed. Both methods result in the milk that was trapped in the milk duct being squirted out the end into a bucket that is supported between the knees (or rests on the ground) of the milker, who usually sits on a low stool.

Traditionally the cow, or cows, would stand in the field or paddock while being milked. Young stock, heifers, would have to be trained to remain still to be milked. In many countries, the cows were tethered to a post and milked. The problem with this method is that it relies on quiet, tractable beasts, because the hind end of the cow is not restrained.

In 1937, it was found that bovine somatotropin (BST or bovine growth hormone) would increase the yield of milk. Monsanto Company developed a synthetic (recombinant) version of this hormone (rBST). In February 1994, rBST was approved by the Food and Drug Administration (FDA) for use in the U.S. It was common in the U.S., but has lost popularity due to consumer demands for rBST-free cows. Only about 25% of dairy cows receive rBST anymore.

However, there are claims that this practice can have negative consequences for the animals themselves. A European Union scientific commission was asked to report on the incidence of mastitis and other disorders in dairy cows, and on other aspects of the welfare of dairy cows. The commission's statement, subsequently adopted by the European Union, stated that the use of rBST substantially increased health problems with cows, including foot problems, mastitis and injection site reactions, impinged on the welfare of the animals and caused reproductive disorders. The report concluded that on the basis of the health and welfare of the animals, rBST should not be used. Health Canada prohibited the sale of rBST in 1999; the recommendations of external committees were that, despite not finding a significant health risk to humans, the drug presented a threat to animal health and, for this reason, could not be sold in Canada.

Structure of the Industry

While most countries produce their own milk products, the structure of the dairy industry varies in different parts of the world. In major milk-producing countries most milk is distributed through whole sale markets. In Ireland and Australia, for example, farmers' co-operatives own many of the large-scale processors, while in the United States many farmers and processors do business through individual contracts. In the United States, the country's 196 farmers' cooperatives sold 86% of milk in the U.S. in 2002, with five cooperatives accounting for half that. This was down from 2,300 cooperatives in the 1940s. In developing countries, the past practice of farmers marketing milk in their own neighborhoods is changing rapidly. Notable developments include considerable foreign investment in the dairy industry and a growing role for dairy cooperatives. Output of milk is growing rapidly in such countries and presents a major source of income growth for many farmers.

Wawa Dairy Farms in Pennsylvania

As in many other branches of the food industry, dairy processing in the major dairy producing countries has become increasingly concentrated, with fewer but larger and more efficient plants operated by fewer workers. This is notably the case in the United States, Europe, Australia and New Zealand. In 2009, charges of anti-trust violations have been made against major dairy industry players in the United States, which critics call Big Milk. Another round of price fixing charges was settled in 2016.

Government intervention in milk markets was common in the 20th century. A limited anti-trust exemption was created for U.S. dairy cooperatives by the Capper-Volstead Act of 1922. In the 1930s, some U.S. states adopted price controls, and Federal Milk Marketing Orders started under the Agricultural Marketing Agreement Act of 1937 and continue in the 2000s. The Federal Milk Price Support Program began in 1949. The Northeast Dairy Compact regulated wholesale milk prices in New England from 1997 to 2001.

Plants producing liquid milk and products with short shelf life, such as yogurts, creams and soft cheeses, tend to be located on the outskirts of urban centres close to consumer markets. Plants manufacturing items with longer shelf life, such as butter, milk powders, cheese and whey powders, tend to be situated in rural areas closer to the milk supply. Most large processing plants tend to specialise in a limited range of products. Exceptionally, however, large plants producing a wide range of products are still common in Eastern Europe, a holdover from the former centralized, supply-driven concept of the market under Communist governments.

As processing plants grow fewer and larger, they tend to acquire bigger, more automated and more efficient equipment. While this technological tendency keeps manufacturing costs lower, the need for long-distance transportation often increases the environmental impact.

Milk production is irregular, depending on cow biology. Producers must adjust the mix of milk which is sold in liquid form vs. processed foods (such as butter and cheese) depending on changing supply and demand.

Farming

A rotary dairy shed

When it became necessary to milk larger cows, the cows would be brought to a shed or barn that was set up with bails (milking stalls) where the cows could be confined while they were milked. One person could milk more cows this way, as many as 20 for a skilled worker. But having cows standing about in the yard and shed waiting to be milked is not good for the cow, as she needs as much time in the paddock grazing as is possible. It is usual to restrict the twice-daily milking to a maximum of an hour and a half each time. It makes no difference whether one milks 10 or 1000 cows, the milking time should not exceed a total of about three hours each day for any cow.

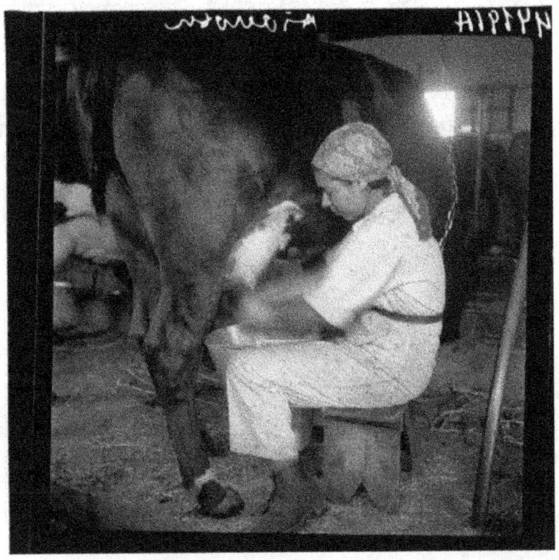

A cow being milked in Israel

As herd sizes increased there was more need to have efficient milking machines, sheds, milk-storage facilities (vats), bulk-milk transport and shed cleaning capabilities and the means of getting cows from paddock to shed and back.

As herd numbers increased so did the problems of animal health. In New Zealand two approaches to this problem have been used. The first was improved veterinary medicines (and the government regulation of the medicines) that the farmer could use. The other was the creation of *veterinary clubs* where groups of farmers would employ a veterinarian (vet) full-time and share those services throughout the year. It was in the vet's interest to keep the animals healthy and reduce the number of calls from farmers, rather than to ensure that the farmer needed to call for service and pay regularly.

This daily milking routine goes on for about 300 to 320 days per year that the cow stays in milk. Some small herds are milked once a day for about the last 20 days of the production cycle but this is not usual for large herds. If a cow is left unmilked just once she is likely to reduce milk-production almost immediately and the rest of the season may see her *dried off* (giving no milk) and still consuming feed. However, once-a-day milking is now being practised more widely in New Zealand for profit and lifestyle reasons. This is effective because the fall in milk yield is at least partially offset by labour and cost savings from milking once per day. This compares to some intensive farm systems in the United States that milk three or more times per day due to higher milk yields per cow and lower marginal labor costs.

Farmers who are contracted to supply liquid milk for human consumption (as opposed to milk for processing into butter, cheese, and so on—see milk) often have to manage their herd so that the contracted number of cows are in milk the year round, or the required minimum milk output is maintained. This is done by mating cows outside their natural mating time so that the period when each cow in the herd is giving maximum production is in rotation throughout the year.

Northern hemisphere farmers who keep cows in barns almost all the year usually manage their herds to give continuous production of milk so that they get paid all year round. In the southern hemisphere the cooperative dairying systems allow for two months on no productivity because their systems are designed to take advantage of maximum grass and milk production in the spring and because the milk processing plants pay bonuses in the dry (winter) season to carry the farmers through the mid-winter break from milking. It also means that cows have a rest from milk production when they are most heavily pregnant. Some year-round milk farms are penalised financially for over-production at any time in the year by being unable to sell their overproduction at current prices.

Artificial insemination (AI) is common in all high-production herds.

Industrial Processing

A Fonterra cooperative dairy factory in Australia

Dairy plants process the raw milk they receive from farmers so as to extend its marketable life. Two main types of processes are employed: heat treatment to ensure the safety of milk for human consumption and to lengthen its shelf-life, and dehydrating dairy products such as butter, hard cheese and milk powders so that they can be stored.

Interior of a cheese factory in Seine-et-Marne, France

Cream and Butter

Today, milk is separated by huge machines in bulk into cream and skim milk. The cream is processed to produce various consumer products, depending on its thickness, its suitability for culinary uses and consumer demand, which differs from place to place and country to country.

Some cream is dried and powdered, some is condensed (by evaporation) mixed with varying amounts of sugar and canned. Most cream from New Zealand and Australian factories is made into butter. This is done by churning the cream until the fat globules coagulate and form a monolithic mass. This butter mass is washed and, sometimes, salted to improve keeping qualities. The residual buttermilk goes on to further processing. The butter is packaged (25 to 50 kg boxes) and chilled for storage and sale. At a later stage these packages are broken down into home-consumption sized packs.

Skimmed Milk

The product left after the cream is removed is called skim, or skimmed, milk. To make a consumable liquid a portion of cream is returned to the skim milk to make *low fat milk* (semi-skimmed) for human consumption. By varying the amount of cream returned, producers can make a variety of low-fat milks to suit their local market. Other products, such as calcium, vitamin D, and flavouring, are also added to appeal to consumers.

Casein

Casein is the predominant phosphoprotein found in fresh milk. It has a very wide range of uses from being a filler for human foods, such as in ice cream, to the manufacture of products such as fabric, adhesives, and plastics.

Cheese

Cheese is another product made from milk. Whole milk is reacted to form curds that can be compressed, processed and stored to form cheese. In countries where milk is legally allowed to be processed without pasteurization, a wide range of cheeses can be made using the bacteria naturally in the milk. In most other countries, the range of cheeses is smaller and the use of artificial cheese curing is greater. Whey is also the byproduct of this process. Some people with lactose intolerance are surprisingly able to eat certain types of cheese. This is because some traditionally made hard cheeses, and soft ripened cheeses may create less reaction than the equivalent amount of milk because of the processes involved. Fermentation and higher fat content contribute to lesser amounts of lactose. Traditionally made Emmental or Cheddar might contain 10% of the lactose found in whole milk. In addition, the aging methods of traditional cheeses (sometimes over two years) reduce their lactose content to practically nothing. Commercial cheeses, however, are often manufactured by processes that do not have the same lactose-reducing properties. Ageing of some cheeses is governed by regulations; in other cases there is no quantitative indication of degree of ageing and concomitant lactose reduction, and lactose content is not usually indicated on labels.

Whey

In earlier times, whey or milk serum was considered to be a waste product and it was, mostly, fed to pigs as a convenient means of disposal. Beginning about 1950, and mostly since about 1980, lactose and many other products, mainly food additives, are made from both casein and cheese whey.

Yogurt

Yogurt (or yoghurt) making is a process similar to cheese making, only the process is arrested before the curd becomes very hard.

Milk Powders

Milk is also processed by various drying processes into powders. Whole milk, skim milk, buttermilk, and whey products are dried into a powder form and used for human and animal consumption. The main difference between production of powders for human or for animal consumption is in the protection of the process and the product from contamination. Some people drink milk reconstituted from powdered milk, because milk is about 88% water and it is much cheaper to transport the dried product.

Other Milk Products

Kumis is produced commercially in Central Asia. Although it is traditionally made from mare's milk, modern industrial variants may use cow's milk instead.

Milking

Originally, milking and processing took place on the dairy farm itself. Later, cream was separated from the milk by machine on the farm, and transported to a factory to be made into butter. The skim milk was fed to pigs. This allowed for the high cost of transport (taking the smallest volume high-value product), primitive trucks and the poor quality of roads. Only farms close to factories could afford to take whole milk, which was essential for cheesemaking in industrial quantities, to them.

Preserved Express Dairies three-axle Milk Tank Wagon at the Didcot Railway Centre, based on an SR chassis

Originally milk was distributed in 'pails', a lidded bucket with a handle. These proved impractical for transport by road or rail, and so the milk churn was introduced, based on the tall conical shape of the butter churn. Later large railway containers, such as the British Railway Milk Tank Wagon were introduced, enabling the transport of larger quantities of milk, and over longer distances.

The development of refrigeration and better road transport, in the late 1950s, has meant that most farmers milk their cows and only temporarily store the milk in large refrigerated bulk tanks, from where it is later transported by truck to central processing facilities.

Milk churns on a railway platform

In many European countries, particularly the United Kingdom, milk is then delivered direct to customers' homes by a milk float.

Milking Machines

Milking machines are used to harvest milk from cows when manual milking becomes inefficient or labour-intensive. One early model was patented in 1907. The milking unit is the portion of a milking machine for removing milk from an udder. It is made up of a claw, four teatcups, (Shells and rubber liners) long milk tube, long pulsation tube, and a pulsator. The claw is an assembly that connects the short pulse tubes and short milk tubes from the teatcups to the long pulse tube and long milk tube. (Cluster assembly) Claws are commonly made of stainless steel or plastic or both. Teatcups are composed of a rigid outer shell (stainless steel or plastic) that holds a soft inner liner or *inflation*. Transparent sections in the shell may allow viewing of liner collapse and milk flow. The annular space between the shell and liner is called the pulse chamber.

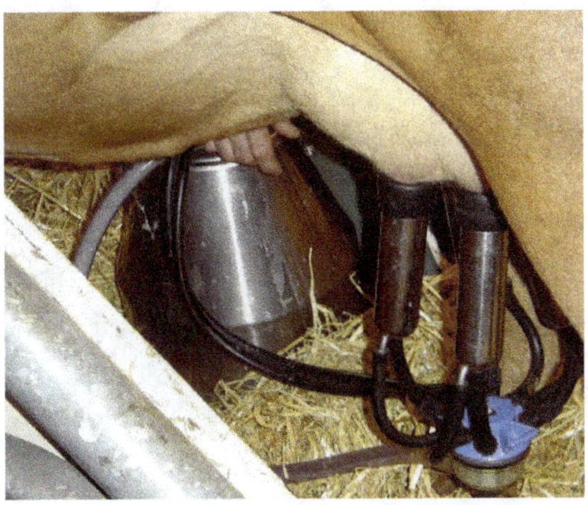

The milking machine extracts milk from all teats.

Milking machines work in a way that is different from hand milking or calf suckling. Continuous vacuum is applied inside the soft liner to massage milk from the teat by creating a pressure difference across the teat canal (or opening at the end of the teat). Vacuum also helps keep the machine attached to the cow. The vacuum applied to the teat causes congestion of teat tissues (accumulation of blood and other fluids). Atmospheric air is admitted into the pulsation chamber about once per second (the pulsation rate) to allow the liner to collapse around the end of teat and relieve congestion in the teat tissue. The ratio of the time that the liner is open (milking phase) and closed (rest phase) is called the pulsation ratio.

The four streams of milk from the teatcups are usually combined in the claw and transported to the milkline, or the collection bucket (usually sized to the output of one cow) in a single milk hose. Milk is then transported (manually in buckets) or with a combination of airflow and mechanical pump to a central storage vat or bulk tank. Milk is refrigerated on the farm in most countries either by passing through a heat-exchanger or in the bulk tank, or both.

The photo to the right shows a bucket milking system with the stainless steel bucket visible on the far side of the cow. The two rigid stainless steel teatcup shells applied to the front two quarters of the udder are visible. The top of the flexible liner is visible at the top of the shells as are the short milk tubes and short pulsation tubes extending from the bottom of the shells to the claw. The bottom of the claw is transparent to allow observation of milk flow. When milking is completed the vacuum to the milking unit is shut off and the teatcups are removed.

Milking machines keep the milk enclosed and safe from external contamination. The interior 'milk contact' surfaces of the machine are kept clean by a manual or automated washing procedures implemented after milking is completed. Milk contact surfaces must comply with regulations requiring food-grade materials (typically stainless steel and special plastics and rubber compounds) and are easily cleaned.

Most milking machines are powered by electricity but, in case of electrical failure, there can be an alternative means of motive power, often an internal combustion engine, for the vacuum and milk pumps.

Milking Shed Layouts

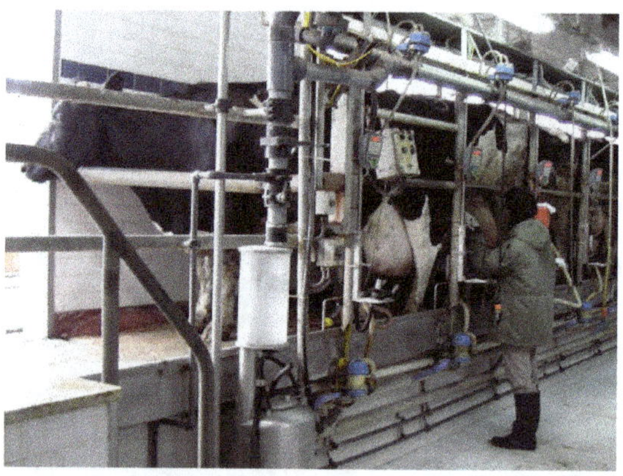

Milking parlour at Pardes Hanna Agricultural High School, Israel

Bail-style Sheds

This type of milking facility was the first development, after open-paddock milking, for many farmers. The building was a long, narrow, *lean-to* shed that was open along one long side. The cows were held in a yard at the open side and when they were about to be milked they were positioned in one of the bails (stalls). Usually the cows were restrained in the bail with a breech chain and a rope to restrain the outer back leg. The cow could not move about excessively and the milker could expect not to be kicked or trampled while sitting on a (three-legged) stool and milking into a bucket. When each cow was finished she backed out into the yard again. The UK bail, initially developed by Wiltshire dairy farmer Arthur Hosier, was a six standing mobile shed with steps that the cow mounted, so the herdsman didn't have to bend so low. The milking equipment was much as today, a vacuum from a pump, pulsators, a claw-piece with pipes leading to the four shells and liners that stimulate and suck the milk from the teat. The milk went into churns, via a cooler.

As herd sizes increased a door was set into the front of each bail so that when the milking was done for any cow the milker could, after undoing the leg-rope and with a remote link, open the door and allow her to exit to the pasture. The door was closed, the next cow walked into the bail and was secured. When milking machines were introduced bails were set in pairs so that a cow was being milked in one paired bail while the other could be prepared for milking. When one was finished the machine's cups are swapped to the other cow. This is the same as for *Swingover Milking Parlours* as described below except that the cups are loaded on the udder from the side. As herd numbers increased it was easier to double-up the cup-sets and milk both cows simultaneously than to increase the number of bails. About 50 cows an hour can be milked in a shed with 8 bails by one person. Using the same teat cups for successive cows has the danger of transmitting infection, mastitis, from one cow to another. Some farmers have devised their own ways to disinfect the clusters between cows.

Herringbone Milking Parlours

In herringbone milking sheds, or parlours, cows enter, in single file, and line up almost perpendicular to the central aisle of the milking parlour on both sides of a central pit in which the milker works (you can visualise a fishbone with the ribs representing the cows and the spine being the milker's working area; the cows face outward). After washing the udder and teats the cups of the milking machine are applied to the cows, from the rear of their hind legs, on both sides of the working area. Large herringbone sheds can milk up to 600 cows efficiently with two people.

Swingover Milking Parlours

Swingover parlours are the same as herringbone parlours except they have only one set of milking cups to be shared between the two rows of cows, as one side is being milked the cows on the other side are moved out and replaced with unmilked ones. The advantage of this system is that it is less costly to equip, however it operates at slightly better than half-speed and one would not normally try to milk more than about 100 cows with one person.

Rotary Milking Sheds

Rotary milking sheds (also known as Rotary milking parlor) consist of a turntable with about 12 to 100 individual stalls for cows around the outer edge. A "good" rotary will be operated with 24–32 (~48–50+) stalls by one (two) milkers. The turntable is turned by an electric-motor drive at a rate that one turn is the time for a cow to be milked completely. As an empty stall passes the entrance a cow steps on, facing the center, and rotates with the turntable. The next cow moves into the next vacant stall and so on. The operator, or milker, cleans the teats, attaches the cups and does any other feeding or whatever husbanding operations that are necessary. Cows are milked as the platform rotates. The milker, or an automatic device, removes the milking machine cups and the cow backs out and leaves at an exit just before the entrance. The rotary system is capable of milking very large herds—over a thousand cows.

Automatic Milking Sheds

Automatic milking or 'robotic milking' sheds can be seen in Australia, New Zealand, the U.S., Canada, and many European countries. Current automatic milking sheds use the voluntary milking (VM) method. These allow the cows to voluntarily present themselves for milking at any time of the day or night, although repeat visits may be limited by the farmer through computer software. A robot arm is used to clean teats and apply milking equipment, while automated gates direct cow traffic, eliminating the need for the farmer to be present during the process. The entire process is computer controlled.

Supplementary Accessories in Sheds

Farmers soon realised that a milking shed was a good place to feed cows supplementary foods that overcame local dietary deficiencies or added to the cows' wellbeing and production. Each bail might have a box into which such feed is delivered as the cow arrives so that she is eating while being milked. A computer can read the eartag of each animal to ration the correct individual supplement. A close alternative is to use 'out-of-parlour-feeders', stalls that respond to a transponder around the cow's neck that is programmed to provide each cow with a supplementary feed, the quantity dependent on her production, stage in lactation, and the benefits of the main ration

The holding yard at the entrance of the shed is important as a means of keeping cows moving into the shed. Most yards have a powered gate that ensures that the cows are kept close to the shed.

Water is a vital commodity on a dairy farm: cows drink about 20 gallons (80 litres) a day, sheds need water to cool and clean them. Pumps and reservoirs are common at milking facilities. Water can be warmed by heat transfer with milk.

Temporary Milk Storage

Milk coming from the cow is transported to a nearby storage vessel by the airflow leaking around the cups on the cow or by a special "air inlet" (5-10 l/min free air) in the claw. From there it is pumped by a mechanical pump and cooled by a heat exchanger. The milk is then stored in a large vat, or bulk tank, which is usually refrigerated until collection for processing.

Waste Disposal

In countries where cows are grazed outside year-round, there is little waste disposal to deal with. The most concentrated waste is at the milking shed, where the animal waste may be liquefied (during the water-washing process) or left in a more solid form, either to be returned to be used on farm ground as organic fertilizer.

In the associated milk processing factories, most of the waste is washing water that is treated, usually by composting, and spread on farm fields in either liquid or solid form. This is much different from half a century ago, when the main products were butter, cheese and casein, and the rest of the milk had to be disposed of as waste (sometimes as animal feed).

Manure spreader going to the field from a dairy farm, Elba, New York.

In dairy-intensive areas, various methods have been proposed for disposing of large quantities of milk. Large application rates of milk onto land, or disposing in a hole, is problematic as the residue from the decomposing milk will block the soil pores and thereby reduce the water infiltration rate through the soil profile. As recovery of this effect can take time, any land based application needs to be well managed and considered. Other waste milk disposal methods commonly employed include solidification and disposal at a solid waste landfill, disposal at a wastewater treatment plant, or discharge into a sanitary sewer.

Associated Diseases

Dairy products manufactured under unsanitary or unsuitable conditions have an increased chance of containing bacteria. Proper sanitation practices help to reduce the rate of bacterial contamination, and pasteurization greatly decreases the amount of contaminated milk that reaches the consumer. Many countries have required government oversight and regulations regarding dairy production, including requirements for pasteurization.

- Leptospirosis is an infection that can be transmitted to people who work in dairy production through exposure to urine or to contaminated water or soil.
- Cowpox is a virus that today is rarely found in either cows or humans. It is a historically important disease, as it led to the first vaccination against the now eradicated smallpox.
- Tuberculosis is able to be transmitted from cattle mainly via milk products that are unpasteurised. The disease has been eradicated from many countries by testing for the disease and culling suspected animals.
- Brucellosis is a bacterial disease transmitted to humans by dairy products and direct animal contact. Brucellosis has been eradicated from certain countries by testing for the disease and culling suspected animals.
- Listeria is a bacterial disease associated with unpasteurised milk, and can affect some cheeses made in traditional ways. Careful observance of the traditional cheesemaking methods achieves reasonable protection for the consumer.

- Crohn's disease has been linked to infection with the bacterium *M. paratuberculosis*, which has been found in pasteurized retail milk in the UK and the USA. *M. paratuberculosis* causes a similar disorder, Johne's disease, in livestock.

Animal Welfare

A portion of the population, including many vegans and Jains, object to dairy production as unethical, cruel to animals, and environmentally deleterious. They do not consume dairy products. They state that cattle suffer under conditions employed by the dairy industry.

Confectionery

Confectionery, also called sweets or candy, is sweet food. The term varies among English-speaking countries. In general, though, confectionery is divided into two broad and somewhat overlapping categories, bakers' confections and sugar confections.

This Kransekake is a traditional Scandinavian baker's confection.

Bakers' confectionery, also called flour confections, includes principally sweet pastries, cakes, and similar baked goods. In the Middle East and Asia, flour-based confections are more dominant.

Sugar confectionery includes sweets, candied nuts, chocolates, chewing gum, sweetmeats, pastillage, and other confections that are made primarily of sugar. In some cases, chocolate confections (confections made of chocolate) are treated as a separate category, as are sugar-free versions of sugar confections. The words *candy* (US and Canada), *sweets* (UK and Ireland), and *lollies* (Australia and New Zealand) are common words for the most common varieties of sugar confectionery.

The confectionery industry also includes specialized training schools and extensive historical records. Traditional confectionery goes back to ancient times, and continued to be eaten through the Middle Ages into the modern era.

Nutrition

Generally, confections are low in micronutrients and protein but high in calories. They may be fat-free foods, although some confections, especially fried doughs, are high-fat foods. Many confections are considered empty calories.

Specially formulated chocolate has been manufactured in the past for military use as a high-density food energy source.

Sweetening Agents

Confections are defined by the presence of sweeteners. These are usually sugars, but it is possible to buy sugar-free sweets, such as sugar-free peppermints. The most common sweetener for home cooking is table sugar, which is chemically a disaccharide called *sucrose*. Hydrolysis of sucrose gives a mixture called invert sugar, which is sweeter and is also a common commercial ingredient. Finally confections, especially commercial ones, are sweetened by a variety of syrups obtained by hydrolysis of starch. These sweeteners include all types of corn syrup.

Bakers' Confectionery

Bakers' confectionery includes sweet baked goods, especially those that are served for the dessert course. Bakers' confections are sweet foods that feature flour as a main ingredient and are baked. Major categories include cakes, sweet pastries, doughnuts, scones, and cookies.

Petits fours are baker's confections.

Sugar Confectionery

Sugar confections include sweet, sugar-based foods, which are usually eaten as snack food. This includes sugar candies, chocolates, candied fruits and nuts, chewing gum, and sometimes ice cream.

The United Nations' International Standard Industrial Classification of All Economic Activities (ISIC) scheme (revision 4) classifies both chocolate and sugar confectionery as ISIC 1073, which includes the manufacture of chocolate and chocolate confectionery; sugar confectionery proper (caramels, cachous, nougats, fondant, white chocolate), chewing gum, preserving fruit, nuts, fruit peels, and making confectionery lozenges and pastilles. In the European Union, the Statistical Classification of Economic Activities in the European Community (NACE) scheme (revision 2) matches the UN classification, under code number 10.82. In the United States, the North American Industry Classification System (NAICS 2012) splits sugar confectionery across three categories: National industry code 311340 for all non-chocolate confectionery manufacturing, 311351 for chocolate and confectionery manufacturing from cacao beans, and national industry 311352 for confectionery manufacturing from purchased chocolate. Ice cream and sorbet are classified with dairy products under ISIC 1050, NACE 10.52, and NAICS 311520.

Different dialects of English use regional terms for sugar confections:

- In Britain, Ireland, and some Commonwealth countries, *sweets* or, more colloquially, *sweeties* (particularly used by children, the Scottish Gaelic word *suiteis* is a derivative).
- In Australia and New Zealand, *lollies*. *Chewy* and *Chuddy* are Australian slang for chewing gum.
- In North America, *candy*, although this term generally refers to a specific range of confectionery and does not include some items of sugar confectionery (e.g. ice cream). *Sweet* is occasionally used, as well as *treat*.

Examples

Sugar confectionery items include sweets, lollipops, candy bars, chocolate, cotton candy, and other sweet items of snack food. Some of the categories and types of sugar confectionery include the following:

Rock candy is simply sugar with coloring or flavor added.

Brittles are a combination of nuts and caramelized sugar.

Chocolate is a common and popular confectionery and can be used in a wide variety of ways.

- Caramels: Derived from a mixture of sucrose, glucose syrup, and milk products. The mixture does not crystallize, thus remains tacky.
- Chocolates: Bite-sized confectioneries generally made with chocolate.
- Divinity: A nougat-like confectionery based on egg whites with chopped nuts.
- Dodol: A toffee-like food delicacy popular in Indonesia, Malaysia, and the Philippines
- Dragée: Sugar-coated almonds and other types of sugar panned candy.
- Fondant: Prepared from a warm mixture of glucose syrup and sucrose, which is partially crystallized. The fineness of the crystallites results in a creamy texture.
- Fudge: Made by boiling milk and sugar to the soft-ball stage. In the US, it tends to be chocolate-flavored.
- Halvah: Confectionery based on tahini, a paste made from ground sesame seeds.
- Hard candy: Based on sugars cooked to the hard-crack stage. Examples include suckers (known as *boiled sweets* in British English), lollipops, jawbreakers (or gobstoppers), lemon drops, peppermint drops and disks, candy canes, rock candy, etc. Also included are types often mixed with nuts such as brittle. Others contain flavorings including coffee such as Kopiko.
- Ice cream: Frozen, flavoured cream, often containing small pieces of chocolate, fruits and/ or nuts.
- Jelly candies: Including those based on sugar and starch, pectin, gum, or gelatin such as Turkish delight (lokum), jelly beans, gumdrops, jujubes, gummies, etc.
- Liquorice: Containing extract of the liquorice root. Chewier and more resilient than gum/ gelatin candies, but still designed for swallowing. For example, Liquorice allsorts. Has a similar taste to star anise.
- Marshmallow: "Peeps" (a trade name), circus peanuts, fluffy puff, Jet-Puffed Marshmallows etc.
- Marzipan: An almond-based confection, doughy in consistency, served in several different ways.
- Mithai: A generic term for confectionery in India, typically made from dairy products and/ or some form of flour. Sugar or molasses are used as sweeteners.

- Tablet: A crumbly milk-based soft and hard candy, based on sugars cooked to the soft ball stage. Comes in several forms, such as wafers and heart shapes.
- Taffy or chews: A candy that is folded many times above 120 °F (50 °C), incorporating air bubbles thus reducing its density and making it opaque.

Cultural Roles

A Japanese vendor selling sweets in "The Great Buddha Sweet Shop" from the *Miyako meisho zue* (1787)

Both bakers' and sugar confections are used to offer hospitality to guests.

Confections are used to mark celebrations or events, such as a wedding cake, birthday cake or Halloween.

Tourists commonly eat confections as part of their travels. The indulgence in rich, sugary foods is seen as a special treat, and choosing local specialties is popular. For example, visitors to Vienna eat Sachertorte and visitors to seaside resorts in the UK eat Blackpool rock candy. Transportable confections like fudges and tablet may be purchased as souvenirs.

Risks

Excessive consumption of confectionery has been associated with increased incidences of type 2 diabetes, obesity, and tooth decay.

Contaminants and coloring agents in confectionery can be particularly harmful to children. Therefore, confectionery contaminants, such as high levels of lead, have been restricted to 1 ppm in the US. There is no specific maximum in the EU.

Candy colorants, particularly yellow colorants such as E102 Tartrazine, E104 Quinoline Yellow and E110 Sunset Yellow FCF, do have many restrictions around the world. Tartrazine, for example, can cause allergic and asthmatic reactions and was once banned in Austria, Germany, and Norway. Some countries such as the UK have asked the food industry to phase out the use of these colorants, especially for products marketed to children.

Candy Making

Candy making is the preparation of candies and sugar confections. Candy making includes the preparation of many various candies, such as hard candies, jelly beans, gumdrops, taffy, liquorice, cotton candy, chocolates and chocolate truffles, dragées, fudge, caramel candy and toffee.

Candy is made by dissolving sugar in water or milk to form a syrup, which is boiled until it reaches the desired concentration or starts to caramelize. The type of candy depends on the ingredients and how long the mixture is boiled. Candy comes in a wide variety of textures, from soft and chewy to hard and brittle. A chocolatier is a person who prepares confectionery from chocolate, and is distinct from a chocolate maker, who create chocolate from cacao beans and other ingredients. Cotton candy is a form of spun sugar often prepared using a cotton candy machine.

Candy being panned (coated) in a giant pot at a candy factory in Nablus, West Bank

Coconut candy being prepared in the Mekong delta area, Vietnam

Hot liquid candy being poured into candy molds by a candymaker

History

The technology for candy making has generally kept pace with the technology of the times. For example, when steam power became common in factories, steam power was also used in candy factories.

Candy making and consumption increased greatly during the Industrial Revolution in the 19th century. Candy had previously been made by hand, either occasionally at home or by specialists in small, local businesses. Increased mechanization caused prices to drop and production to increase.

In the late 19th century and especially the early 20th century, industrial candy making was almost exclusively a masculine affair, and home-based candy making was a feminine affair. Candy was considered sweet and dainty, so making it at home, giving it away to friends, and perhaps selling small amounts in the local area, conformed with the Western gender roles for women of the time. Most women making and selling candy did so only seasonally or for a little extra money; they rarely earned enough to support themselves or their families. Despite several large brands being named after women or otherwise capitalizing on wholesome, feminine, and maternal images, very few were owned or operated by women.

Safety

Making candy can be hazardous due to the use of boiled sugar and melted chocolate. Boiling sugar often exceeds 150 °C (302 °F)—hotter than most cooked foods—and the sugar tends to stick to the skin, causing burns and blisters upon skin contact. Worker safety programs focus on reducing contact between workers and hot food or hot equipment, and reducing splashing, because even small splashes can cause burns. Some ingredients can also irritate the eyes and lungs, if, for example, powdered ingredients are accidentally inhaled, so worker protection involves reducing exposure to potentially irritating ingredients.

Hard Candy

Hard candy, also referred to as boiled sweet, is a candy prepared from one or more syrups boiled to a temperature of 160 °C (320 °F). After a syrup boiled to this temperature cools, it is called hard candy, since it becomes stiff and brittle as it approaches room temperature. Hard candy recipes variously call for syrups of sucrose, glucose, or fructose. To add color, food coloring is sometimes used.

Fruit-shaped hard candy

Sugar Stages

The final texture of candy depends on the sugar concentration. As the syrup is heated, it boils and water evaporates as the sugar concentration increases. As a result, the boiling point rises because it is a colligative property and related to the concentration of the sugar solution. A given temperature corresponds to a particular sugar concentration because of this physical correlation, so temperature is used as a marker for the necessary concentration. In general, higher temperatures and greater sugar concentrations result in hard, brittle candies, and lower temperatures result in softer candies. The stages of sugar cooking are as follows:

Stage	Temperature	Sugar concentration
thread (e.g., syrup)	110 to 112 °C (230 to 234 °F)	80%
soft ball (e.g., fudge)	112 to 116 °C (234 to 241 °F)	85%
firm ball (e.g., soft caramel candy)	118 to 120 °C (244 to 248 °F)	87%
hard ball (e.g., nougat)	121 to 130 °C (250 to 266 °F)	90%
soft crack (e.g., salt water taffy)	132 to 143 °C (270 to 289 °F)	95%
hard crack (e.g., toffee)	146 to 154 °C (295 to 309 °F)	99%
clear liquid	160 °C (320 °F)	100%
brown liquid (e.g., liquid caramel)	170 °C (338 °F)	100%
burnt sugar	177 °C (351 °F)	100%

The names come from the methods used to test the syrup before thermometers became affordable. The "thread" stage is tested by cooling a little syrup, and pulling it between the thumb and forefinger. When the correct stage is reached, a thread will form. This stage is used for making syrups. For subsequent stages, a small spoonful of syrup is dropped into cold water, and the characteristics of the resulting lump are evaluated to determine the concentration of the syrup. A smooth lump indicates "ball" stages, with the corresponding hardness described. At the "soft crack" stage, the syrup forms threads that are just pliable. At the "hard crack" stage, the threads are brittle.

This method is still used today in some kitchens. A candy thermometer is more convenient, but has the drawback of not automatically adjusting for local conditions such as altitude, as the cold water test does.

Once the syrup reaches 171 °C (340 °F) or higher, the sucrose molecules break down into many simpler sugars, creating an amber-colored substance known as caramel. This should not be confused with caramel candy, although it is the candy's main flavoring.

Soft Candy

Cotton Candy

Cotton candy, also known as Candy Floss, is a form of spun sugar. Typical machines used to make cotton candy include a spinning head enclosing a small bowl into which granulated sugar is poured. Colored sugar or separate sugar and food coloring are used to provide color. Heaters near the rim of the head melt the sugar, which is squeezed out through tiny holes by centrifugal force, and the molten sugar solidifies in the air and is caught in a larger bowl which totally surrounds the spinning head. After the product builds up on the inside walls of the larger bowl, a stick, cone, or hands are used inserted, upon which the sugar strands are gathered.

The bowl of a cotton candy machine

- Cotton candy being prepared

The spinning head of a cotton candy machine

Marshmallows

Marshmallows are prepared using egg whites, corn syrup and sugar. The use of marshmallow to make a sweet dates back to ancient Egypt, where the recipe called for extracting sap from the marshmallow plant (*Althaea officinalis*) and mixing it with nuts and honey. Another pre-modern recipe uses the pith of the marshmallow plant, rather than the sap. In modern times, marshmallows are often commercially prepared using extrusion.

Marshmallow creme being prepared

Chocolatiering

Chocolatiering involves the preparing of confections from chocolate. A chocolatier that has mastered the artistry of chocolate may be referred to as a Master Chocolatier. Chocolatiering involves the techniques of tempering, molding and sculpting. Tempering is a heat treatment method performed on chocolate involving heating and cooling the chocolate to result in desired characteristics like shininess of the chocolate or 'snap', the way it breaks. Molding is a design technique used in making chocolate pieces that are of a certain shape by taking liquid chocolate and pouring it into a mold and letting it harden. Sculpting is a type of three-dimensional artwork that may involve using molds and pieces of chocolate, and decorating the piece with designs in chocolate.

A chocolatier making chocolate eggs

A chocolatier making a chocolate tower

Tools and Machinery

A variety of tools and machines are used in making candy, ranging from simple kitchen tools like bowls and spoons to elaborate factory machinery.

Because exact temperature control is critical for some candies, a common tool is the candy thermometer. Inexpensive candy thermometers measure food temperatures up to about 160 °C, and those designed for commercial candy production may run even higher.

A starch mogul is used in candy factories to shape soft candies or candy centers from syrups or gels. These centers may then be sent through a chocolate enrober to coat them in chocolate.

Baking

Baking is a method of cooking food that uses prolonged dry heat, normally in an oven, but also in hot ashes, or on hot stones. The most common baked item is bread but many other types of foods are baked. Heat is gradually transferred "from the surface of cakes, cookies, and breads to their centre. As heat travels through it transforms batters and doughs into baked goods with a firm dry crust and a softer centre". Baking can be combined with grilling to produce a hybrid barbecue variant by using both methods simultaneously, or one after the other. Baking is related to barbecuing because the concept of the masonry oven is similar to that of a smoke pit.

Bedouin Arabs making and baking bread

Because of historical social and familial roles, baking has traditionally been performed at home by women for domestic consumption and by men in bakeries and restaurants for local consumption. When production was industrialized, baking was automated by machines in large factories. The art of baking remains a fundamental skill and is important for nutrition, as baked goods, especially breads, are a common but important food, both from an economic and cultural point of view. A person who prepares baked goods as a profession is called a baker.

Anders Zorn - *Bread baking* (1889)

Foods and Techniques

All types of food can be baked, but some require special care and protection from direct heat. Various techniques have been developed to provide this protection.

A Palestinian woman baking *markook* bread on tava or Saj oven in Artas, Bethlehem

In addition to bread, baking is used to prepare cakes, pastries, pies, tarts, quiches, cookies, scones, crackers, pretzels, and more. These popular items are known collectively as "baked goods," and are often sold at a bakery, which is a store that carries only baked goods, or at markets, grocery stores, or through other venues.

A terracotta baking mould for pastry or bread, representing goats and a lion attacking a cow. Early 2nd millennium BC, Royal palace at Mari, Syria

Meat, including cured meats, such as ham can also be baked, but baking is usually reserved for meatloaf, smaller cuts of whole meats, or whole meats that contain stuffing or coating such as bread crumbs or buttermilk batter. Some foods are surrounded with moisture during baking by placing a small amount of liquid (such as water or broth) in the bottom of a closed pan, and letting it steam up around the food, a method commonly known as braising or slow baking. Larger cuts prepared without stuffing or coating are more often roasted, which is a similar process, using higher temperatures and shorter cooking times. Roasting, however, is only suitable for finer cuts of meat, so other methods have been developed to make tougher meat cuts palatable after baking. One of these is the method known as *en croûte* (French for "in a crust"), which protects the food from direct heat and seals the natural juices inside. Meat, poultry, game, fish or vegetables can be prepared by baking *en croûte*. Well-known examples include Beef Wellington, where the beef is encased in pastry before baking; pâté en croûte, where the terrine is encased in pastry before baking; and the Vietnamese variant, a meat-filled pastry called pâté chaud. The *en croûte* method also allows meat to be baked by burying it in the embers of a fire – a favourite method of cooking venison. In this case, the protective casing (or crust) is made from a paste of flour and water and is discarded before eating. Salt can also be used to make a protective crust that is not eaten. Another method of protecting food from the heat while it is baking, is to cook it *en papillote* (French for "in parchment"). In this method, the food is covered by baking paper (or aluminium foil) to protect it while it is being baked. The cooked parcel of food is sometimes served unopened, allowing diners to discover the contents for themselves which adds an element of surprise.

Eggs can also be used in baking to produce savoury or sweet dishes. In combination with dairy products especially cheese, they are often prepared as a dessert. For example, although a baked custard can be made using starch (in the form of flour, cornflour, arrowroot, or potato flour), the flavour of the dish is much more delicate if eggs are used as the thickening agent. Baked custards, such as crème caramel, are among the items that need protection from an oven's direct heat, and the *bain-marie* method serves this purpose. The cooking container is half submerged in water in another, larger one, so that the heat in the oven is more gently applied during the baking process. Baking a successful soufflé requires that the baking process be carefully controlled. The oven temperature must be absolutely even and the oven space not shared with another dish. These factors, along with the theatrical effect of an air-filled dessert, have given this baked food a reputation for being a culinary achievement. Similarly, a good baking technique (and a good oven) are also needed to create a baked Alaska because of the difficulty of baking hot meringue and cold ice cream at the same time.

Baking can also be used to prepare various other foods such as pizzas, baked potatoes, baked apples, baked beans, some casseroles and pasta dishes such as lasagne.

Baking in Ancient Times

An Egyptian funerary Model of a bakery and brewery (11th dynasty, circa 2009–1998 B.C.)

The first evidence of baking occurred when humans took wild grass grains, soaked them in water, and mixed everything together, mashing it into a kind of broth-like paste. The paste was cooked by pouring it onto a flat, hot rock, resulting in a bread-like substance. Later, when humans mastered fire, the paste was roasted on hot embers, which made bread-making easier, as it could now be made any time fire was created. The world's oldest oven was discovered in Croatia in 2014 dating back 6500 years ago. The Ancient Egyptians baked bread using yeast, which they had previously been using to brew beer. Bread baking began in Ancient Greece around 600 BC, leading to the invention of enclosed ovens. "Ovens and worktables have been discovered in archaeological digs from Turkey (Hacilar) to Palestine (Jericho) and date back to 5600 BC."

Baking flourished during the Roman Empire. Beginning around 300 BC, the pastry cook became an occupation for Romans (known as the pastillarium) and became a respected profession because pastries were considered decadent, and Romans loved festivity and celebration. Thus, pastries were often cooked especially for large banquets, and any pastry cook who could invent new types of tasty treats was highly prized. Around 1 AD, there were more than three hundred pastry chefs in Rome, and Cato wrote about how they created all sorts of diverse foods and flourished professionally and socially because of their creations. Cato speaks of an enormous number of breads including; libum (sacrificial cakes made with flour), placenta (groats and cress), spira (our modern day flour pretzels), scibilata (tortes), savaillum (sweet cake), and globus apherica (fritters). A great selection of these, with many different variations, different ingredients, and varied patterns, were often found at banquets and dining halls. The Romans baked bread in an oven with its own chimney, and had mills to grind grain into flour. A bakers' guild was established in 168 BC in Rome.

Commercial Baking

Eventually, the Roman art of baking became known throughout Europe and eventually spread to eastern parts of Asia. By the 13th century in London, commercial trading, including baking, had many regulations attached. In the case of food, they were designed to create a system "so there was little possibility of false measures, adulterated food or shoddy manufactures." There were by that time twenty regulations applying to bakers alone, including that every baker had to have "the impression of his seal" upon each loaf of bread.

Vincent van Gogh – The Bakery in Noordstraat (1882)

Beginning in the 19th century, alternative leavening agents became more common, such as baking soda. Bakers often baked goods at home and then sold them in the streets. This scene was so common that Rembrandt, among others, painted a pastry chef selling pancakes in the streets of Germany, with children clamoring for a sample. In London, pastry chefs sold their goods from handcarts. This developed into a delivery system of baked goods to households and greatly increased demand as a result. In Paris, the first open-air café of baked goods was developed, and baking became an established art throughout the entire world.

Every family used to prepare the bread for its own consumption, the *trade* of baking, not having yet taken shape. Mrs Beeton (1861)

Schulze Baking Company Factory (1914–15)

Baking eventually developed into a commercial industry using automated machinery which enabled more goods to be produced for widespread distribution. In the United States, the baking industry "was built on marketing methods used during feudal times and production techniques developed by the Romans." Some makers of snacks such as potato chips or crisps have produced baked versions of their snack products as an alternative to the usual cooking method of deep-frying in an attempt to reduce their calorie or fat content. Baking has opened up doors to businesses such as cake shops and factories where the baking process is done with larger amounts in large, open furnaces.

The aroma and texture of baked goods as they come out of the oven are strongly appealing but is a quality that is quickly lost. Since the flavour and appeal largely depend on freshness, commercial producers have to compensate by using food additives as well as imaginative labeling. As more and more baked goods are purchased from commercial suppliers, producers try to capture that original appeal by adding the label "home-baked." Such attempts seek to make an emotional link to the remembered freshness of baked goods as well as to attach positive associations the purchaser has with the idea of "home" to the bought product. Freshness is such an important quality that restaurants, although they are commercial (and not domestic) preparers of food, bake their own products. For example, scones at The Ritz London Hotel "are not baked until early afternoon on the day they are to be served, to make sure they are as fresh as possible."

Equipment

Baking needs an enclosed space for heating – typically in an oven. The fuel can be supplied by wood, coal, gas, or electricity. Adding and removing items from an oven may be done by hand with an oven mitt or by a peel, a long handled tool specifically used for that purpose.

Many commercial ovens are equipped with two heating elements: one for baking, using convection and thermal conduction to heat the food, and one for broiling or grilling, heating mainly by radiation. Another piece of equipment still used for baking is the Dutch oven. "Also called a bake kettle, bastable, bread oven, fire pan, bake oven kail pot, tin kitchen, roasting kitchen, *doufeu* (French: "gentle fire") or *feu de compagne* (French: "country oven") [it] originally replaced the cooking jack as the latest fireside cooking technology," combining "the convenience of pot-oven and hangover oven."

Asian cultures have adopted steam baskets to produce the effect of baking while reducing the amount of fat needed.

Process

Baking bread at the Roscheider Hof Open Air Museum

Baked goods

Eleven events occur concurrently during baking, some of which (such as starch glutenization) would not occur at room temperature.

1. Fats melt;
2. Gases form and expand
3. Microorganisms die
4. Sugar dissolves
5. Egg, milk, and gluten proteins coagulate
6. Starches gelatinise
7. Gases evaporate
8. Caramelization and Maillard browning occur on crust
9. Enzymes are inactivated
10. Changes occur to nutrients
11. Pectin breaks down.

The dry heat of baking changes the form of starches in the food and causes its outer surfaces to brown, giving it an attractive appearance and taste. The browning is caused by caramelization of sugars and the Maillard reaction. Maillard browning occurs when "sugars break down in the presence of proteins". Because foods contain many different types of sugars and proteins, Maillard browning contributes to the flavour of a wide range of foods, including nuts, roast beef and baked bread." The moisture is never entirely "sealed in"; over time, an item being baked will become dry. This is often an advantage, especially in situations where drying is the desired outcome, like drying herbs or roasting certain types of vegetables.

The baking process does not require any fat to be used to cook in an oven. When baking, consideration must be given to the amount of fat that is contained in the food item. Higher levels of fat such as margarine, butter, lard, or vegetable shortening will cause an item to spread out during the baking process.

With the passage of time, breads harden and become stale. This is not primarily due to moisture being lost from the baked products, but more a reorganization of the way in which the water and starch are associated over time. This process is similar to recrystallization and is promoted by storage at cool temperatures, such as in a domestic refrigerator or freezer.

Cultural and Religious Significance

Baking, especially of bread, holds special significance for many cultures. It is such a fundamental part of everyday food consumption that the children's nursery rhyme *Pat-a-cake, pat-a-cake, baker's man* takes baking as its subject. Baked goods are normally served at all kinds of party and special attention is given to their quality at formal events. They are also one of the main components of a tea party, including at nursery teas and high teas, a tradition which started in Victorian Britain, reportedly when Anna Russell, Duchess of Bedford "grew tired of the sinking feeling which afflicted her every afternoon round 4 o'clock ... In 1840, she plucked up courage and asked for a tray of tea, bread and butter, and cake to be brought to her room. Once she had formed the habit she found she could not break it, so spread it among her friends instead. As the century progressed, afternoon tea became increasingly elaborate."

Bird baked from bread on the March equinox to celebrate spring and the forty martyrs

Benedictine Sisters of the Benedectine Monastery of Caltanissetta producing the crocette, they used to be prepared for the Holy Crucifix festivity. This was situated next to the Church of the Holy Cross, from which the sweets take the name.

Baking matzot at Kfar Chabad

For Jews, Matzo is a baked product of considerable religious and ritual significance. Baked matzah bread can be ground up and used in other dishes, such as Gefilte fish, and baked again. For Christians, bread has to be baked to be used as an essential component of the sacrament of the Eucharist. In the Eastern Christian tradition, baked bread in the form of birds is given to children to carry to the fields in a spring ceremony that celebrates the Forty Martyrs of Sebaste.

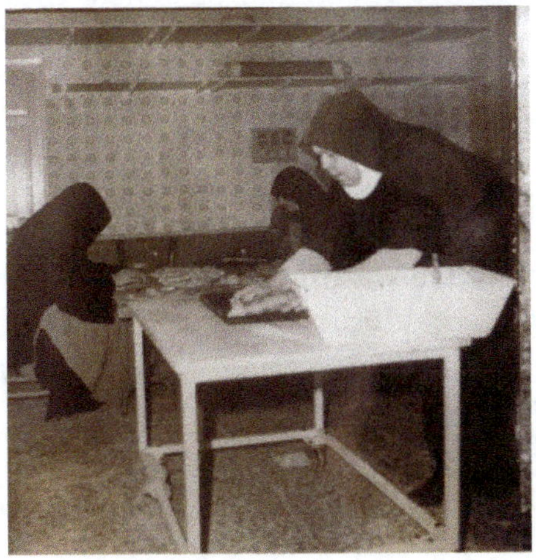

Benedictine Sisters of Caltanissetta producing the crocetta of Caltanissetta

Snack

A snack is a portion of food, smaller than a regular meal, generally eaten between meals. Snacks come in a variety of forms including packaged snack foods and other processed foods, as well as items made from fresh ingredients at home.

A picture of some low-calorie fruit and vegetable snacks, including apples, asparagus, beetroots, bell peppers, endives, and tomatoes.

Traditionally, snacks are prepared from ingredients commonly available in the home. Often cold cuts, fruit, leftovers, nuts, sandwiches, and the like are used as snacks. The Dagwood sandwich was originally the humorous result of a cartoon character's desire for large snacks. With the spread of convenience stores, packaged snack foods became a significant business. Snack foods are typically designed to be portable, quick, and satisfying. Processed snack foods, as one form of convenience food, are designed to be less perishable, more durable, and more portable than prepared foods. They often contain substantial amounts of sweeteners, preservatives, and appealing ingredients such as chocolate, peanuts, and specially-designed flavors (such as flavored potato chips).

Beverages, such as coffee, are not generally considered snacks though they may be consumed along with or in lieu of snack foods.

A snack eaten shortly before going to bed or during the night may be called a midnight snack.

History

In the United States, the first snack food was the peanut. Peanuts first arrived from South America via slave ships, and became incorporated into African-inspired cooking on southern plantations. After the Civil War, the taste for peanuts spread north, where they were incorporated into the culture of baseball games and vaudeville theaters.

Along with popcorn (also of South American origin), snacks bore the stigma of being sold by unhygienic street vendors. The middle-class etiquette of the Victorian era (1837-1901) categorized any food that did not require proper usage of utensils as lower-class.

Pretzels were introduced to North America by the Dutch via New Amsterdam in the 17th century. In the 1860s, the snack was still associated with immigrants, unhygienic street vendors, and saloons. Due to loss of business during the Prohibition era (1920-1933), pretzels underwent rebranding to make them more appealing to the public. Packaging revolutionized snack foods, allowing sellers to reduce contamination risk while making it easy to advertise brands with a logo. Pretzels boomed in popularity, bringing many other types of snack foods with it. By the 1950s, snacking had become an all-American pastime, becoming an internationally recognized emblem of middle American life.

Snacks and Health

Healthy snacks include those that have significant vitamins, are low in saturated fat and added sugars, and have a low sodium content. Examples of healthy snacks include:

- Foods that have whole grains
- Fruits and vegetables
- Nuts and seeds
- Low-fat dairy products
- Lean meats, eggs, such as hard-boiled eggs, and lean cheese

Nutritional Concerns

Government bodies like Health Canada are recommending that people make a conscious effort to eat more healthy, natural snacks – such as fruit, vegetables, nuts and cereal grains – while avoiding high-calorie, low-nutrient junk food.

A 2010 study showed that children in the United States snacked on average six times per day, approximately twice as often as American children in the 1970s. This represents consumption of roughly 570 calories more per day than U.S. children consumed in the 1970s.

Snacks and Cognition

A Tufts University, Department of Psychology empirical study titled, "Effect of an afternoon confectionery snack on cognitive processes critical to learning" found that a consumption of a confectionery snack in the afternoon improved spatial memory in the study's sample group, but in the area of attention performance had a mixed effect.

Types of Snack Foods

- Almonds
- Apple slices
- Antz
- Bagel with cream cheese
- Bitterballen
- Bread/toast with butter, honey, jam, or other spread
- Candy bar
- Carrot Chips
- Cashews
- Cheese puffs/Cheese curls
- Cheese, a larger cold prepared snack
- Chocolate-coated marshmallow treats
- Corn chips and Tortilla chips
- Cocktail sausage
- Crackers
- Cookies/Biscuits
- Deviled eggs
- Doughnuts
- Dried fruits
- Drinkable yogurt
- Edamame, fresh or dried
- Granola bars

- Falafel
- Flour tortilla with a filling
- Frozen berries
- Sliced fruit
- Fruit cocktail
- Fruit salad
- Ice cream
- Instant noodles
- Jell-O
- Jerky
- Kaassoufflé
- Milkshake
- Pound cake, in slices
- Lunchables
- Mixed nuts
- Muffins
- Nachos
- Papadum
- Parsnip bits
- Peanuts
- Pita bread, straight from the packet or toasted
- Pizza
- Popcorn
- Pork pie
- Pork rinds
- Potato chips
- Pakoda
- Pretzels, hard or soft
- Raisins
- Ratatouille, served cold, a larger cold prepared snack

- Rice cake
- Rice crackers, distinguished from the above
- Saltines
- Sandwich, a larger cold prepared snack
- Samosa
- Seeds (sunflower or seed mix)
- Shortbread
- Smoked salmon
- Smoothie
- Teacake
- Toast
- Trail mix
- Vegetables (e.g. carrots, celery, cherry tomatoes)
- Whole fruit
- Yogurt

Candy

Popcorn

A candy bar of the Snickers brand

Trail mix

References

- International Food Information Service, ed. (2009). Dictionary of Food Science and Technology (2nd ed.). Chichester, U.K.: Wiley–Blackwell. p. 106. ISBN 9781405187404.
- International Food Information Service, ed. (2009). Dictionary of Food Science and Technology (2nd ed.). Chichester, U.K.: Wiley–Blackwell. p. 39. ISBN 9781405187404.

- Cleave, Paul (2012). "Sugar in Tourism: 'Wrapped in Devonshire Sunshine'". Sugar Heritage and Tourism in Transition. Channel View Publications. pp. 159–172. ISBN 9781845413897.
- Figoni, Paula I. (2011). How Baking Works: Exploring the Fundamentals of Baking Science (3rd ed.). New Jersey: John Wiley & Sons. ISBN 978-0-470-39813-5.p.38
- Rochelle, Jay Cooper (2001). Bread for the Wilderness: Baking As Spiritual Craft. Fairfax, VA: Xulon Press. p. 32. ISBN 1-931232-52-0.
- Bessie Emrick Whitten (1990). David O. Whitten, ed., Handbook of American Business History: Manufacturing. Connecticut: Greenwood Publishing Group. ISBN 0-313-25198-3.p.53
- Simpson, Helen (1986). The London Ritz Book of Afternoon Tea - The Art & Pleasures of Taking Tea. London, UK: Angus & Robertson, Publishers. p. 8. ISBN 0-207-15415-5.
- Simpson, Helen (1986). The London Ritz Book of Afternoon Tea: The Art & Pleasures of Making Tea. London: Angus & Robertson Publishers. p. 16. ISBN 0-207-15415-5.
- Carroll, Abigail (2013-08-30). "How Snacking Became Respectable". Wall Street Journal. ISSN 0099-9660. Retrieved 2016-05-29.
- ABC News. "American Diet Then and Now: How Snacking Is Expanding the Country's Waistline - ABC News". ABC News. Retrieved 20 February 2016.
- "ISIC Rev.4 code 1073: Manufacture of cocoa, chocolate and sugar confectionery". United Nations Statistics Division, Classification Registry. 2014. Retrieved 18 June 2014.
- Pfister, Fred. "Pfister Consulting: History of Baking – How Did It All Start? Yes people". Archived from the original on 3 February 2013. Retrieved January 1, 2013.

3

Food Additives: Significant Topics

Processed food usually contains a medley of additives in the form of food substitutes, food coloring etc. Food fortification is a method used to increase the nutrient value of food by the addition of vitamins, minerals and proteins with the aim of providing a balanced diet or overcoming dietary deficiencies. This chapter examines these food additives, their purposes, regulations related to them and their types.

Food Additive

Food additives are substances added to food to preserve flavor or enhance its taste and appearance. Some additives have been used for centuries; for example, preserving food by pickling (with vinegar), salting, as with bacon, preserving sweets or using sulfur dioxide as with wines. With the advent of processed foods in the second half of the twentieth century, many more additives have been introduced, of both natural and artificial origin.

Numbering

To regulate these additives, and inform consumers, each additive is assigned a unique number, termed as "E numbers", which is used in Europe for all approved additives. This numbering scheme has now been adopted and extended by the *Codex Alimentarius* Commission to internationally identify all additives, regardless of whether they are approved for use.

E numbers are all prefixed by "E", but countries outside Europe use only the number, whether the additive is approved in Europe or not. For example, acetic acid is written as E260 on products sold in Europe, but is simply known as additive 260 in some countries. Additive 103, alkanet, is not approved for use in Europe so does not have an E number, although it is approved for use in Australia and New Zealand. Since 1987, Australia has had an approved system of labelling for additives in packaged foods. Each food additive has to be named or numbered. The numbers are the same as in Europe, but without the prefix "E".

The United States Food and Drug Administration (FDA) lists these items as "generally recognized as safe" (GRAS); they are listed under both their Chemical Abstracts Service number and FDA regulation under the United States Code of Federal Regulations.

Categories

Food additives can be divided into several groups, although there is some overlap because some additives exert more than one effect. For example, salt is both a preservative as well as a flavor.

Acidulents

confer sour or acid taste. Common acidulents include vinegar, citric acid, tartaric acid, malic acid, fumaric acid, and lactic acid.

Acidity regulators

Acidity regulators are used to control the pH of foods for stability or to affect activity of enzymes.

Anticaking agents

Anticaking agents keep powders such as milk powder from caking or sticking.

Antifoaming and foaming agents

Antifoaming agents reduce or prevent foaming in foods. Foaming agents do the reverse.

Antioxidants

Antioxidants such as vitamin C are preservatives by inhibiting the degradation of food by oxygen.

Bulking agents

Bulking agents such as starch are additives that increase the bulk of a food without affecting its taste.

Food coloring

Colorings are added to food to replace colors lost during preparation or to make food look more attractive.

Color retention agents

In contrast to colorings, color retention agents are used to preserve a food's existing color.

Emulsifiers

Emulsifiers allow water and oils to remain mixed together in an emulsion, as in mayonnaise, ice cream, and homogenized milk.

Flavors

Flavors are additives that give food a particular taste or smell, and may be derived from natural ingredients or created artificially.

Flavor enhancers

Flavor enhancers enhance a food's existing flavors. A popular example is monosodium glutamate. Some flavor enhancers have little flavor independent of the food.

Flour treatment agents

Flour treatment agents are added to flour to improve its color or its use in baking.

Glazing agents

Glazing agents provide a shiny appearance or protective coating to foods.

Humectants

Humectants prevent foods from drying out.

Tracer gas

Tracer gas allow for package integrity testing to prevent foods from being exposed to atmosphere, thus guaranteeing shelf life.

Preservatives

Preservatives prevent or inhibit spoilage of food due to fungi, bacteria and other microorganisms.

Stabilizers

Stabilizers, thickeners and gelling agents, like agar or pectin (used in jam for example) give foods a firmer texture. While they are not true emulsifiers, they help to stabilize emulsions.

Sweeteners

Sweeteners are added to foods for flavoring. Sweeteners other than sugar are added to keep the food energy (calories) low, or because they have beneficial effects for diabetes mellitus and tooth decay and diarrhea.

Thickeners

Thickening agents are substances which, when added to the mixture, increase its viscosity without substantially modifying its other properties.

Safety and Regulation

With the increasing use of processed foods since the 19th century, food additives are more widely used. Many countries regulate their use. For example, boric acid was widely used as a food preservative from the 1870s to the 1920s, but was banned after World War I due to its toxicity, as demonstrated in animal and human studies. During World War II, the urgent need for cheap, available food preservatives led to it being used again, but it was finally banned in the 1950s. Such cases led to a general mistrust of food additives, and an application of the precautionary principle led to the conclusion that only additives that are known to be safe should be used in foods. In the United States, this led to the adoption of the Delaney clause, an amendment to the Federal Food, Drug, and Cosmetic Act of 1938, stating that no carcinogenic substances may be used as food additives. However, after the banning of cyclamates in the United States and Britain in 1969, saccharin, the only remaining legal artificial sweetener at the time, was found to cause cancer in rats. Widespread public outcry in the United States, partly communicated to Congress by postage-paid postcards supplied in the packaging of sweetened soft drinks, led to the retention of saccharin, despite its violation of the Delaney clause. However, in 2000, saccharin was later found to only be carcinogenic to rats due to their unique urine chemistry.

Hyperactivity

Periodically, concerns have been expressed about a linkage between additives and hyperactivty, however "no clear evidence of ADHD was provided".

In 2007, Food Standards Australia New Zealand published an official shoppers' guidance with which the concerns of food additives and their labeling are mediated. In the EU it can take 10 years or more to obtain approval for a new food additive. This includes five years of safety testing, followed by two years for evaluation by the European Food Safety Authority and another three years before the additive receives an EU-wide approval for use in every country in the European Union. Apart from testing and analyzing food products during the whole production process to ensure safety and compliance with regulatory standards, Trading Standards officers (in the UK) protect the public from any illegal use or potentially dangerous mis-use of food additives by performing random testing of food products.

There has been significant controversy associated with the risks and benefits of food additives. Natural additives may be similarly harmful or be the cause of allergic reactions in certain individuals. For example, safrole was used to flavor root beer until it was shown to be carcinogenic. Due to the application of the Delaney clause, it may not be added to foods, even though it occurs naturally in sassafras and sweet basil.

Micronutrients

A subset of food additives, micronutrients added in food fortification processes preserve nutrient value by providing vitamins and minerals to foods such as flour, cereal, margarine and milk which normally would not retain such high levels. Preservatives also reduce spoilage from sources such as air, bacteria, fungi, and yeast.

Standardization of its Derived Products

ISO has published a series of standards regarding the topic and these standards are covered by ICS 67.220.

Science

Many food additives absorb radiation in the ultraviolet and / or visible region of the spectrum. This absorbance can be used to determine the concentration of an additive in a sample using external calibration. However, additives may occur together and the absorbance by one could interfere with the absorbance of another. A prior separation stage is necessary and the additives are first separated by high-pressure liquid chromatography (HPLC) and then determined on-line using a UV and/ or visible light detector.

Food Fortification

Food fortification or enrichment is the process of adding micronutrients (essential trace elements and vitamins) to food. It may be a purely commercial choice to provide extra nutrients in a food, while other times it is a public health policy which aims to reduce the number of people with dietary deficiencies within a population.

Manufacturers once proposed selling fortified junk food and beer, but USFDA policies of the time forbade it

Diets that lack variety can be deficient in certain nutrients. Sometimes the staple foods of a region can lack particular nutrients, due to the soil of the region or because of the inherent inadequacy of the normal diet. Addition of micronutrients to staples and condiments can prevent large-scale deficiency diseases in these cases.

While it is true that both fortification and enrichment refer to the addition of nutrients to food, the true definitions do slightly vary. As defined by the World Health Organization (WHO) and the Food and Agricultural Organization of the United Nations (FAO), fortification refers to "the practice of deliberately increasing the content of an essential micronutrient, ie. vitamins and minerals (including trace elements) in a food irrespective of whether the nutrients were originally in the food before processing or not, so as to improve the nutritional quality of the food supply and to provide a public health benefit with minimal risk to health," whereas enrichment is defined as "synonymous with fortification and refers to the addition of micronutrients to a food which are lost during processing."

Food fortification was identified as the second strategy of four by the WHO and FAO to begin decreasing the incidence of nutrient deficiencies at the global level.

As outlined by the FAO, the most common fortified foods are:

- Cereals and cereal based products
- Milk and Milk products
- Fats and oils

- Accessory food items
- Tea and other beverages
- Infant formulas

Types

The four main methods of food fortification (named as to indicate the procedure that is used in order to fortify the food):

1. Biofortification (i.e. breeding crops to increase their nutritional value, which can include both conventional selective breeding, and modern genetic modification)
2. Synthetic biology (i.e. addition of probiotic bacteria to foods)
3. Commercial and industrial fortification (i.e. flour, rice, oils (common cooking foods))
4. Home fortification (e.g. vitamin D drops)

Rationale

The WHO and FAO, among many other nationally recognized organizations, have recognized that there are over 2 billion people worldwide who suffer from a variety of micronutrient deficiencies. In 1992, 159 countries pledged at the FAO/WHO International Conference on Nutrition to make efforts to help combat these issues of micronutrient deficiencies, highlighting the importance of decreasing the number of those with iodine, vitamin A, and iron deficiencies. A significant statistic that led to these efforts was the discovery that approximately 1 in 3 people worldwide were at risk for either an iodine, vitamin A, or iron deficiency. Although it is recognized that food fortification alone will not combat this deficiency, it is a step towards reducing the prevalence of these deficiencies and their associated health conditions.

In Canada, The Food and Drug Regulations have outlined specific criterion which justifies food fortification:

1. To replace nutrients which were lost during manufacturing of the product (i.e. the manufacturing of flour)
2. To act as a public health intervention
3. To ensure the nutritional equivalence of substitute foods (i.e. to make butter and margarine similar in content, soy milk and cow's milk, etc.)
4. To ensure the appropriate vitamin and mineral nutrient composition of foods for special dietary purposes (i.e., gluten-free products, low sodium, or any other products specifically designed for special dietary requirements from an individual).

There are also several advantages to approaching nutrient deficiencies among populations via food fortification as opposed to other methods. These may include, but are not limited to: treating a population without specific dietary interventions therefore not requiring a change in dietary patterns, continuous delivery of the nutrient, does not require individual compliance, and potential to maintain nutrient stores more efficiently if consumed on a regular basis.

Criticism

Several organizations such as the WHO, FAO, Health Canada, and the Nestlé Research Center acknowledge that there are limitations to food fortification. Within the discussion of nutrient deficiencies the topic of nutrient toxicities can also be immediately questioned. Fortification of nutrients in foods may deliver toxic amounts of nutrients to an individual and also cause its associated side effects. As seen with the case of excessive fluoride intakes below, the result can be irreversible staining to the teeth. Although this may be a minor toxic effect to health, there are several that are more severe.

The WHO states that limitations to food fortification may include: human rights issues indicating that consumers have the right to choose if they want fortified products or not, the potential for insufficient demand of the fortified product, increased production costs leading to increased retail costs, the potential that the fortified products will still not be a solution to nutrient deficiencies amongst low income populations who may not be able to afford the new product, and children who may not be able to consume adequate amounts thereof.

Food safety worries led to legislation in Denmark in 2004 restricting foods fortified with extra vitamins or minerals. Products banned include: Rice Crispies, Shreddies, Horlicks, Ovaltine and Marmite.

Danes said [Kelloggs] Corn Flakes, Rice Krispies and Special K wanted to include "toxic" doses which, if eaten regularly, could damage children's livers and kidneys and harm fetuses in pregnant women.

Limited Absorption

One factor that limits the benefits of food fortification is that isolated nutrients added back into a processed food that has had many of its nutrients removed, does not always result in the added nutrients being as bioavailable as they would be in the original, whole food. An example is skim milk that has had the fat removed, and then had vitamin A and vitamin D added back. Vitamins A and D are both fat-soluble and non-water-soluble, so a person consuming skim milk in the absence of fats may not be able to absorb as much of these vitamins as one would be able to absorb from drinking whole milk. On the other hand, the nutrient added as a fortificant may have a higher bioavailability than from foods, which is the case with folic acid used to increase folate intakes.

Phytochemicals such as polyphenols, phytic acid in cereal grains and tannin in tea can also impact nutrient absorption, limiting the bioavailability of intrinsic and additional nutrients, and reducing the effectiveness of fortification programs.

Excess Intake of Micronutrients

Ecological studies have shown that increased B vitamin fortification is correlated with the prevalence of obesity and diabetes. Daily consumption of iron per capita in the United States has dramatically surged since World War II and nearly doubled over the past century due to increases in iron fortification and increased consumption of meat. Existing evidence suggests that excess iron intake may play a role in the development of obesity, cardiovascular disease, diabetes and cancer.

Fortification of foods with folic acid has been mandated in many countries solely to improve the folate status of pregnant women to prevent neural tube defects—a relatively rare birth defect which affected 0.5% of US births before fortification began. However, when fortification is introduced, several hundred thousand people are exposed to an increased intake of folic acid for each neural tube defect pregnancy that is prevented. In humans, increased folic acid intake leads to elevated blood concentrations of naturally occurring folates and of unmetabolized folic acid. High blood concentrations of folic acid may decrease natural killer cell cytotoxicity, and high folate status may reduce the response to drugs used to treat malaria, rheumatoid arthritis, psoriasis, and cancer. A combination of high folate levels and low vitamin B-12 status may be associated with an increased risk of cognitive impairment and anemia in the elderly and, in pregnant women, with an increased risk of insulin resistance and obesity in their children. Folate has a dual effect on cancer, protecting against cancer initiation but facilitating progression and growth of preneoplastic cells and subclinical cancers. Furthermore, intake of folic acid from fortification have turned out to be significantly greater than originally modeled in pre mandate predictions. Therefore, a high folic acid intake due to fortification may be harmful for more people than the policy is designed to help.

Different Forms of Micronutrients

There is a concern that micronutrients are legally defined in such a way that does not distinguish between different forms, and that fortified foods often have nutrients in a balance that would not occur naturally. For example, in the U.S., food is fortified with folic acid, which is one of the many naturally-occurring forms of folate, and which only contributes a minor amount to the folates occurring in natural foods. In many cases, such as with folate, it is an open question of whether or not there are any benefits or risks to consuming folic acid in this form.

In many cases, the micronutrients added to foods in fortification are synthetic.

In some cases, certain forms of micronutrients can be actively toxic in a sufficiently high dose, even if other forms are safe at the same or much higher doses. There are examples of such toxicity in both synthetic and naturally-occurring forms of vitamins. Retinol, the active form of Vitamin A, is toxic in a much lower dose than other forms, such as beta carotene. Menadione, a phased-out synthetic form of Vitamin K, is also known to be toxic.

Food Supplements

There are several main groups of food supplements like:

- Vitamins and co-vitamins
- Essential minerals
- Essential fatty acids
- Essential amino acids

- Phytonutrients
- Enzymes

Examples of Fortification in Foods

Many foods and beverages worldwide have been fortified, whether a voluntary action by the product developers or by law. Although some may view these additions as strategic marketing schemes to sell their product, there is a lot of work that must go into a product before simply fortifying it. In order to fortify a product, it must first be proven that the addition of this vitamin or mineral is beneficial to health, safe, and an effective method of delivery. The addition must also abide by all food and labeling regulations and support nutritional rationale. From a food developer's point of view, they also need to consider the costs associated with this new product and whether or not there will be a market to support the change.

Examples of foods and beverages that have been fortified and shown to have positive health effects:

Iodized Salt

"Iodine deficiency disorder (IDD) is the single greatest cause of preventable mental retardation. Severe deficiencies cause cretinism, stillbirth and miscarriage. But even mild deficiency can significantly affect the learning ability of populations........ Today over 1 billion people in the world suffer from iodine deficiency, and 38 million babies born every year are not protected from brain damage due to IDD."—Kul Gautam, Deputy Executive Director, UNICEF, October 2007

Iodised salt has been used in the United States since before World War II. It was discovered in 1821 that goiters could be treated by the use of iodized salts. However, it was not until 1916 that the use of iodized salts could be tested in a research trial as a preventative measure against goiters. By 1924, it became readily available in the US.

Currently in Canada and the US, the RDA for iodine is as low as 90 μg/day for children (4–8 years) and as high as 290 μg/day for breast-feeding mothers.

Diseases that are associated with an iodine deficiency include: mental retardation, hypothyroidism, and goiter. There is also a risk of various other growth and developmental abnormalities.

Folic Acid

Folic acid (also known as folate) functions in reducing blood homocysteine levels, forming red blood cells, proper growth and division of cells, and preventing neural tube defects (NTDs).

In many industrialized countries, the addition of folic acid to flour has prevented a significant number of NTDs in infants. Two common types of NTDs, spina bifida and anencephaly, affect approximately 2500-3000 infants born in the US annually. Research trials have shown the ability to reduce the incidence of NTDs by supplementing pregnant mothers with folic acid by 72%.

The RDA for folic acid ranges from as low as 150 μg/day for children aged 1–3 years old, to 400 μg/day for males and females over the age of 19, and 600 μg/day during pregnancy.

Diseases associated with folic acid deficiency include: megaloblastic or macrocytic anemia, cardiovascular disease, certain types of cancer, and NTDs in infants.

Niacin

Niacin has been added to bread in the USA since 1938 (when voluntary addition started), a programme which substantially reduced the incidence of pellagra. As early as 1755, pellagra was recognized by doctors as being a niacin deficiency disease. Although not officially receiving its name of pellagra until 1771. Pellagra was seen amongst poor families who used corn as their main dietary staple. Although corn itself does contain niacin, it is not a bioavailable form unless it undergoes Nixtamalization (treatment with alkali, traditional in Native American cultures) and therefore was not contributing to the overall intake of niacin. Although pellagra can still be seen in developing countries, fortification of food with niacin played a huge role in eliminating the prevalence of the disease.

The RDA for niacin is 2 mg NE(niacin equivalents)/day (AI) for infants aged 0–6 months, 16 mg NE/day for males, and 14 mg NE/day for females who are over the age of 19.

Diseases associated with niacin deficiency include: Pellagra which consisted of signs and symptoms called the 3D's-"Dermatitis, dementia, and diarrhea. Others may include vascular or gastrointestinal diseases.

Common diseases which present a high frequency of niacin deficiency: alcoholism, anorexia nervosa, HIV infection, gastrectomy, malabsorptive disorders, certain cancers and their associated treatments.

Vitamin D

Since Vitamin D is a fat-soluble vitamin, it cannot be added to a wide variety of foods. Foods that it is commonly added to are margarine, vegetable oils and dairy products. During the late 1800s, after the discovery of curing conditions of scurvy and beriberi had occurred, researchers were aiming to see if the disease, later known as rickets, could also be cured by food. Their results showed that sunlight exposure and cod liver oil were the cure. It was not until the 1930s that vitamin D was actually linked to curing rickets. This discovery led to the fortification of common foods such as milk, margarine, and breakfast cereals. This took the astonishing statistics of approximately 80–90% of children showing varying degrees of bone deformations due to vitamin D deficiency to being a very rare condition.

Risk factors for vitamin D deficiencies include:

- In infants, being exclusively or primarily breast-fed
- Dark skin
- Living in cold climates and having little sun exposure
- Being elderly
- Covering all or almost all of one's skin while outdoors

- Liberal use of high-SPF sunscreens
- Fat malabsorption syndromes
- Inflammatory bowel diseases
- Obesity

The current RDA for infants aged 0–6 months is 10 μg (400 International Units (IU))/day and for adults over 19 years of age it is 15 μg (600 IU)/day.

Diseases associated with a vitamin D deficiency include rickets, osteoporosis, and certain types of cancer (breast, prostate, colon and ovaries). It has also been associated with increased risks for fractures, heart disease, type 2 diabetes, autoimmune and infectious diseases, asthma and other wheezing disorders, myocardial infarction, hypertension, congestive heart failure, and peripheral vascular disease.

Fluoride

Although fluoride is not considered an essential mineral, it is seen as possibly useful in prevention of tooth decay and maintaining adequate dental health. In the mid-1900s it was discovered that towns with a high level of fluoride in their water supply was causing the residents' teeth to have both brown spotting and a strange resistance to dental caries. This led to the fortification of water supplies with fluoride with safe amounts to retain the properties of resistance to dental caries but avoid the staining cause by fluorosis (a condition caused by excessive fluoride intake). The tolerable upper intake level (UL) set for fluoride ranges from 0.7 mg/day for infants aged 0–6 months and 10 mg/day for adults over the age of 19.

Conditions commonly associated with fluoride deficiency are dental caries and osteoporosis.

Others

Some other examples of fortified foods:

- Calcium is frequently added to fruit juices, carbonated beverages and rice.
- White rice is frequently enriched to replace some of the lost nutrients during milling or adding extras in.
- "Golden rice" is a variety of rice which has been genetically modified to produce beta carotene.
- Amylase rich flour is utilized for food making to increase dietary consumption.

Fortification for Body Building

Despite having some scientific basis, but with controversial ethics, is the science of using foods and food supplements to achieve a defined health goal. A common example of this use of food supplements is the extent to which body builders will use amino acid mixtures, vitamins and phytochemicals to enhance natural hormone production, increase muscle and reduce fat. The literature is not concrete on an appropriate method for use of fortification for body builders and therefore may not be recommended due to safety concerns.

Fortification for Medical Treatment

There is interest in the use of food supplements in established medical conditions. This nutritional supplementation using foods as medicine (nutraceuticals) has been effectively used in treating disorders affecting the immune system up to and including cancers. This goes beyond the definition of "food supplement", but should be included for the sake of completeness.

Food Coloring

Food coloring, or color additive, is any dye, pigment or substance that imparts color when it is added to food or drink. They come in many forms consisting of liquids, powders, gels, and pastes. Food coloring is used both in commercial food production and in domestic cooking. Due to its safety and general availability, food coloring is also used in a variety of non-food applications including cosmetics, pharmaceuticals, home craft projects, and medical devices.

The addition of food coloring, such as beta-carotene, gives margarine its yellow color.

Purpose of Food Coloring

People associate certain colors with certain flavors, and the color of food can influence the perceived flavor in anything from candy to wine. Sometimes the aim is to simulate a color that is perceived by the consumer as natural, such as adding red coloring to glacé cherries (which would otherwise be beige), but sometimes it is for effect, like the green ketchup that Heinz launched in 1999. Color additives are used in foods for many reasons including:

- Offset color loss due to exposure to light, air, temperature extremes, moisture and storage conditions
- Correct natural variations in color
- Enhance colors that occur naturally
- Provide color to colorless and "fun" foods

- Make food more attractive and appetizing, and informative
- Allow consumers to identify products on sight, like candy flavors or medicine dosages
- Food colorants, synthetic

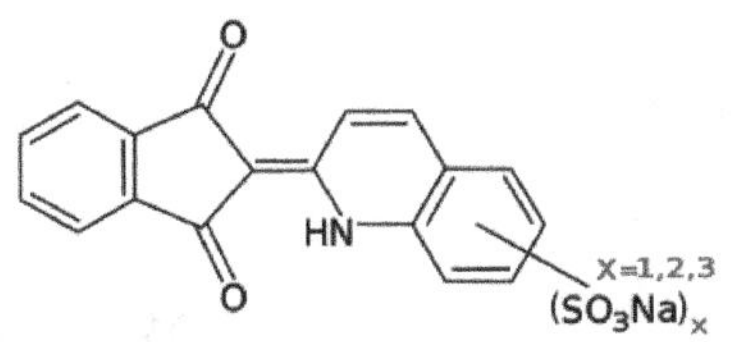

Quinoline Yellow WS, which is yellow.

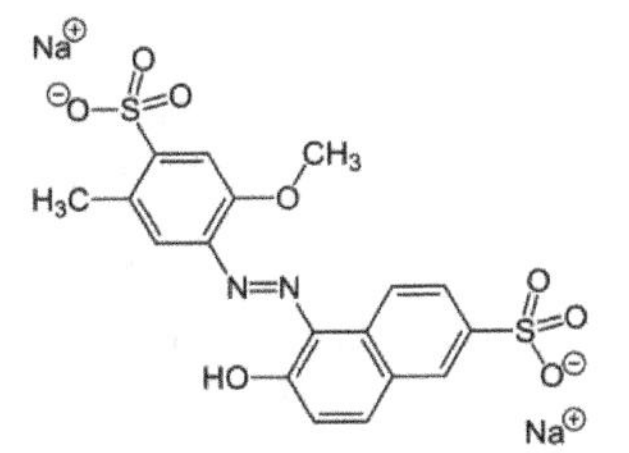

Allura Red AC, which is red.

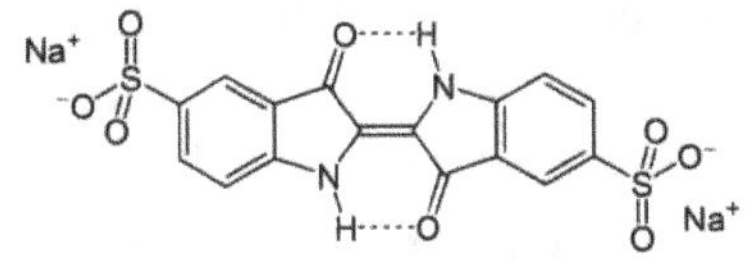

Indigo carmine, which is blue.

- Food colorants, natural

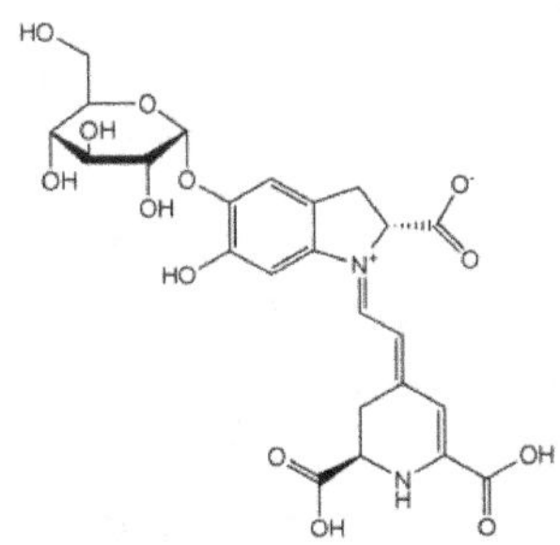

Betanin, a beet-derived dye.

Regulation

Past Use and Regulation History

The addition of colorants to foods is thought to have occurred in Egyptian cities as early as 1500 BC, when candy makers added natural extracts and wine to improve the products' appearance. During the Middle Ages, the economy in the European countries was based on agriculture, and the peasants were accustomed to producing their own food locally or trading within the village communities. Under feudalism, aesthetic aspects were not considered, at least not by the vast majority of the generally very poor population. This situation changed with urbanization at the beginning of the Modern Age, when trade emerged—especially the import of precious spices and colors. One of the very first food laws, created in Augsburg, Germany, in 1531, concerned spices or colorants and required saffron counterfeiters to be burned.

With the onset of the industrial revolution, people became dependent on foods produced by others. These new urban dwellers demanded food at low cost. Analytical chemistry was still primitive and regulations few. The adulteration of foods flourished. Heavy metal and other inorganic element-containing compounds turned out to be cheap and suitable to "restore" the color of watered-down milk and other foodstuffs, some more lurid examples being:

- Red lead (Pb_3O_4) and vermillion (HgS) were routinely used to color cheese and confectionery.

- Copper arsenite ($CuHAsO_3$) was used to recolor used tea leaves for resale. It also caused two deaths when used to color a dessert in 1860.

Sellers at the time offered more than 80 artificial coloring agents, some invented for dyeing textiles, not foods.

...Thus, with potted meat, fish and sauces taken at breakfast he would consume more or less Armenian bole, red lead, or even bisulphuret of mercury. At dinner with his curry or cayenne he would run the chance of a second dose of lead or mercury; with pickles, bottled fruit and vegetables he would be nearly sure to have copper administrated to him; and while he partook of bon-bons at dessert, there was no telling of the number of poisonous pigments he might consume. Again his tea if mixed or green, he would certainly not escape without the administration of a little Prussian blue...

Many color additives had never been tested for toxicity or other adverse effects. Historical records show that injuries, even deaths, resulted from tainted colorants. In 1851, about 200 people were poisoned in England, 17 of them fatally, directly as a result of eating adulterated lozenges. In 1856, mauveine, the first synthetic color, was developed by Sir William Henry Perkin and by the turn of the century, unmonitored color additives had spread through Europe and the United States in all sorts of popular foods, including ketchup, mustard, jellies, and wine.

Concerns over food safety led to numerous regulations throughout the world,. German food regulations released in 1882 stipulated the exclusion of dangerous minerals such as arsenic, copper, chromium, lead, mercury and zinc, which were frequently used as ingredients in colorants. In contrast to today, these first laws followed the principle of a negative listing (substances not allowed for use); they were already driven by the main principles of today's food regulations all over the world, since all of these regulations follow the same goal: the protection of consumers from toxic substances and from fraud. In the United States, the Pure Food and Drug Act of 1906 reduced the permitted list of synthetic colors from 700 down to seven. Even with updated food laws, adulteration continued for many years and this, together with more recent adverse press comments on food colors and health, has continued to contribute to consumer concern about color addition to foodstuffs.

In the 20th century, the improvement of chemical analysis and the development of trials to identify the toxic features of substances added to foods led to the replacement of the negative lists by lists of substances allowed to be used for the production and the improvement of foods. This principle is called a positive listing, and almost all recent legislations are based on it. Positive listing implies that substances meant for human consumption have been tested for their safety, and that they have to meet specified purity criteria prior to their approval by the corresponding authorities. In 1962, the first EU directive (62/2645/EEC) approved 36 colorants, of which 20 were naturally derived and 16 were synthetic. This directive did not list which food products the colorants could or could not be used in. At that time, each member state could designate where certain colors could and could not be used. In Germany, for example, quinoline yellow was allowed in puddings and desserts, but tartrazine was not. The reverse was true in France. This was updated in 1989 with 89/107/EEC, which concerned food additives authorized for use in foodstuffs.

Current Regulation

While naturally derived colors are not required to be certified by a number of regulatory bodies throughout the world (including the U.S. FDA), they still need to be approved for use in that country. The FDA lists "color additives exempt from certification" for food in subpart A of the Code of Federal Regulations - Title 21 Part 73. However, this list contains substances which may have synthetic origins, such as nature identical beta-carotene. FDA's permitted colors are classified as subject to certification or exempt from certification, both of which are subject to rigorous safety standards prior to their approval and listing for use in foods.

- Certified colors are synthetically produced and are used widely because they impart an intense, uniform color, are less expensive, and blend more easily to create a variety of hues. There are nine certified color additives approved for use in the United States. Certified food colors generally do not add undesirable flavors to foods.
- Colors that are exempt from certification include pigments derived from natural sources such as vegetables, minerals or animals. Nature derived color additives are typically more expensive than certified colors and may add unintended flavors to foods. Examples of exempt colors include annatto, beet extract, caramel, beta-carotene, turmeric and grape skin extract.

Food colorings are tested for safety by various bodies around the world and sometimes different bodies have different views on food color safety. In the United States, FD&C numbers (which indicate that the FDA has approved the colorant for use in foods, drugs and cosmetics) are given to approved synthetic food dyes that do not exist in nature, while in the European Union, E numbers are used for all additives, both synthetic and natural, that are approved in food applications. The food colors are known by E numbers that begin with a 1, such as E100 (turmeric) or E161b (lutein). The safety of food colors and other food additives in the EU is evaluated by the European Food Safety Authority. Color Directive 94/36/EC, enacted by the European Commission in 1994, outlines permitted natural and artificial colors with their approved applications and limits in different foodstuffs. This is binding to all member countries of the EU. Any changes have to be implemented into their national laws within a given time frame. In non-EU member states, food additives are regulated by their national authorities, which usually, but not in all cases, try to harmonize with the laws adopted by the EU. Most other countries have their own regulations and list of food colors which can be used in various applications, including maximum daily intake limits.

Global Harmonization

Global trade requires harmonization of food regulations on a world-wide basis in order to abolish barriers of trade and to ensure that the economic and nutritional demands of all nations are considered. Since the beginning of the 1960s, JECFA has played an important role in the development of international standards for food additives, not only by its toxicological assessments, which are continuously published by the WHO in a "Technical Report Series", but furthermore by elaborating appropriate purity criteria, which are laid down in the two volumes of the "Compendium of Food Additive Specifications" and their supplements. These specifications are not legally binding but very often serve as a guiding principle, especially in countries where no scientific expert committees have been established.

In order to further regulate the use of these evaluated additives, in 1962 the WHO and FAO created an international commission, the Codex Alimentarius, which is composed of authorities, food industry associations and consumer groups from all over the world. Within the Codex organization, the Codex Committee for Food Additives and Contaminants is responsible for working out recommendations for the application of food additives, the General Standard for Food Additives. In the light of the World Trade Organizations General Agreement on Tariffs and Trade (GATT), the Codex Standard, although not legally binding, influences food color regulations all over the world.

Artificial Coloring

As the 1900s began, a host of synthetic dyes and pigments became available. Originally, these were dubbed `coal-tar' colors because the starting materials were obtained from bituminous coal. Many synthesized dyes were easier and less costly to produce and were superior in coloring properties when compared to naturally derived alternatives of the time.

Food coloring spreading on a thin water film in the International Space Station

Also known as "azo-dyes", synthetic colors are generally produced via a two-step chemical synthesis. The first step forms a diazo compound from the reaction of aromatic amines generally formed from nitrosamine and a diazonium compound. The second step couples these diazo compounds with various reactive aromatic hydrocarbons. Due to the π-electrons across the two aromatic sections and the azo-groups, a conjugated system exists that is able to absorb light of specific wavelengths, leading to the color of the compounds (this principle also applies to naturally derived pigments as well).

The attractiveness of the synthetic dyes is that their color, lipophilicity, and other attributes can be engineered by the design of the specific molecule. The color of the dyes can be controlled selecting the number of azo-groups and various substituents. Yellow shades are achieved by using aceto-acetanilide and heterocyclic compounds. Red colors result from the reaction between an aniline derivative (diazo) with a naphthol derivate. A blue results from replacing the aniline derivate with a benzidine derivate. The pair indigo and indigo carmine exhibit the same color, but the former is soluble in lipids, and the latter is water soluble because it has been fitted with sulfonate functional groups.

In The United States

Only seven dyes were initially approved under the U.S. Pure Food and Drug Act of 1906, but several have been delisted and replacements have been found. Some of the food colorings have the abbreviation "FCF" in their names. This stands for "For Coloring Food" (US) or "For Colouring of Food" (UK).

In the United States, manufacturers must get FDA approval for every batch of certified colors produced. FDA personnel evaluate the color's physical appearance and chemically analyze it. Tests include purity (total color content), moisture, residual salts, unreacted intermediates, colored impurities other than the main color (called subsidiary colors), any other specified impurities, and heavy metals (lead, arsenic, and mercury). The results are reviewed for compliance with the identity and specifications described in the listing regulation for the color additive. If the sample is found to meet these requirements, the FDA issues a certificate for the batch that identifies the color additive. For example, a batch of "tartrazine" becomes "FD&C Yellow No. 5."

Current Seven Generally Permitted

In the US, the following seven artificial colorings are generally permitted in food (the most common in bold) as of 2016. The lakes of these colorings are also permitted except the lake of Red No. 3.

- FD&C Blue No. 1 – Brilliant Blue FCF, E133 (blue shade)
- FD&C Blue No. 2 – Indigotine, E132 (indigo shade)
- FD&C Green No. 3 – Fast Green FCF, E143 (turquoise shade)
- FD&C Red No. 3 – Erythrosine, E127 (pink shade, commonly used in glacé cherries)
- FD&C Red No. 40 – Allura Red AC, E129 (red shade)
- FD&C Yellow No. 5 – Tartrazine, E102 (yellow shade)
- FD&C Yellow No. 6 – Sunset Yellow FCF, E110 (orange shade)

Permitted for Limited Use in Foods

Two dyes are allowed by the FDA for limited applications:

- Citrus Red 2 (orange shade) - allowed only to color orange peels.
- Orange B (red shade) - allowed only for use in hot dog and sausage casings (not produced after 1978, but never delisted)

Delisted and Banned in The US

- FD&C Red No. 2 – Amaranth, E123
- FD&C Red No. 4
- FD&C Red No. 32 was used to color Florida oranges.
- FD&C Orange Number 1 was one of the first water-soluble dyes to be commercialized, and one of seven original food dyes allowed under the Pure Food and Drug Act of June 30, 1906.
- FD&C Orange No. 2 was used to color Florida oranges.

- FD&C Yellow No. 1, 2, 3, and 4
- FD&C Violet No. 1

Approved in EU

E numbers 102-143 cover the range of artificial colors. For an overview of currently allowed additives see here . Some artificial dyes approved for food use in the EU include:

- E104: Quinoline Yellow
- E122: Carmoisine
- E124: Ponceau 4R
- E131: Patent Blue V
- E142: Green S

Natural Food Dyes

Carotenoids (E160, E161, E164), chlorophyllin (E140, E141), anthocyanins (E163), and betanin (E162) comprise four main categories of plant pigments grown to color food products. Other colorants or specialized derivatives of these core groups include:

Natural food colors can make a variety of different hues

- Annatto (E160b), a reddish-orange dye made from the seed of the achiote
- Caramel coloring (E150a-d), made from caramelized sugar
- Carmine (E120), a red dye derived from the cochineal insect, *Dactylopius coccus*
- Elderberry juice
- Lycopene (E160d)
- Paprika (E160c)
- Turmeric (E100)

Blue colors are especially rare.

To ensure reproducibility, the colored components of these substances are often provided in highly purified form. For stability and convenience, they can be formulated in suitable carrier materials (solid and liquids). Hexane, acetone, and other solvents break down cell walls in the fruit and vegetables and allow for maximum extraction of the coloring. Traces of these may still remain in the finished colorant, but they do not need to be declared on the product label. These solvents are known as carry-over ingredients.

Dyes and Lakes

Color additives are available for use in food as either "dyes" or lake pigments (commonly known as "lakes").

Dyes dissolve in water, but are not soluble in oil. Dyes are manufactured as powders, granules, liquids or other special purpose forms. They can be used in beverages, dry mixes, baked goods, confections, dairy products, pet foods, and a variety of other products. Dyes also have side effects that lakes lack. Consuming large amounts of dyes can color stools.

Lakes are made by combining dyes with salts (usually aluminum salts) to make insoluble compounds. Lakes tint by dispersion. Lakes are not oil-soluble, but are oil-dispersible. Lakes are more stable than dyes and are ideal for coloring products containing fats and oils or items lacking sufficient moisture to dissolve dyes. Typical uses include coated tablets, cake and doughnut mixes, hard candies and chewing gums, lipsticks, soaps, shampoos, talc, etc.

Other Uses

Because food dyes are generally safer to use than normal artists' dyes and pigments, some artists have used food coloring as a means of making pictures, especially in forms such as body-painting. Red food dye is often used in theatrical blood.

Most artificial food colorings are a type of acid dye, and can be used to dye protein fibers and nylon with the addition of an acid. They are all washfast and most are also lightfast. They will not permanently bond to plant fibers and other synthetics.

Criticism and Health Implications

There is no evidence to support broad claims that food coloring causes food intolerance and ADHD-like behavior in children. It is possible that certain food coloring may act as a trigger in those who are genetically predisposed, but the evidence is weak.

After concerns were expressed that food colorings may cause ADHD-like behavior in children, the collective evidence do not support this assertion. The US FDA and other food safety authorities to regularly review the scientific literature, and led the UK Food Standards Agency (FSA) to commission a study by researchers at Southampton University of the effect of a mixture of six food dyes (Tartrazine, Allura Red, Ponceau 4R, Quinoline Yellow WS, Sunset Yellow and Carmoisine (dubbed the "Southampton 6")) on children in the general population. These colorants are found in beverages. The study found "a possible link between the consumption of these artificial colours and a sodium benzoate preservative and increased hyperactivity" in the children; the advisory committee to the FSA that evaluated the study also determined that because of study limitations, the results could not be extrapolated to the general population, and further testing was recommended". The U.S. FDA did not make changes following the publication of the Southampton study, but following a citizen petition filed by the Center for Science in the Public Interest in 2008, requesting the FDA ban several food additives, the FDA reviewed the available evidence, and still made no changes.

The European regulatory community, with an emphasis on the precautionary principle, required labelling and temporarily reduced the acceptable daily intake (ADI) for the food colorings; the UK FSA called for voluntary withdrawal of the colorings by food manufacturers. However, in 2009 the EFSA re-evaluated the data at hand and determined that "the available scientific evidence does not substantiate a link between the color additives and behavioral effects" for any of the dyes.

Sugar Substitute

A sugar substitute is a food additive that provides a sweet taste like that of sugar while containing significantly less food energy. Some sugar substitutes are produced by nature, and others produced synthetically. Those that are not produced by nature are, in general, called artificial sweeteners.

Packets of Assugrin, a brand of cyclamate.

Types of Sugar Substitutes

An important class of sugar substitutes is known as *high-intensity sweeteners*. These are compounds with many times the sweetness of sucrose, common table sugar. As a result, much less sweetener is required and energy contribution is often negligible. The sensation of sweetness caused by these compounds (the "sweetness profile") is sometimes notably different from sucrose, so they are often used in complex mixtures that achieve the most natural sweet sensation.

If the sucrose (or other sugar) that is replaced has contributed to the texture of the product, then a bulking agent is often also needed. This may be seen in soft drinks or sweet teas that are labeled as "diet" or "light" that contain artificial sweeteners and often have notably different mouthfeel, or in table sugar replacements that mix maltodextrins with an intense sweetener to achieve satisfactory texture sensation.

In the United States, seven intensely sweet sugar substitutes have been approved for use. They are stevia, aspartame, sucralose, neotame, acesulfame potassium (Ace-K), saccharin, and advantame. Cyclamates are used outside the U.S., but have been prohibited in the U.S. since 1969.

Others, which may or may not be approved depending on jurisdiction, include allulose (psicose) and monk fruit. The U.S. Food and Drug Administration regulates artificial sweeteners as food additives. Food additives must be approved by the FDA, which publishes a generally recognized as safe (GRAS) list of additives. Stevia is exempt under the FDA's GRAS policy due to its being a natural substance in wide use well before 1958, and the FDA has approved it on these grounds. The conclusions about safety are based on a detailed review of a large body of information, including hundreds of toxicological and clinical studies.

The majority of sugar substitutes approved for food use are artificially synthesized compounds. However, some bulk natural sugar substitutes are known, including sorbitol and xylitol, which are found in berries, fruit, vegetables, and mushrooms. It is not commercially viable to extract these products from fruits and vegetables, so they are produced by catalytic hydrogenation of the appropriate reducing sugar. For example, xylose is converted to xylitol, lactose to lactitol, and glucose to sorbitol. Other natural substitutes are known, but these have yet to gain official approval for food use.

Sorbitol and xylitol are examples of sugar alcohols (also known as polyols). These are, in general, less sweet than sucrose but have similar bulk properties and can be used in a wide range of food products. Sometimes the sweetness profile is fine-tuned by mixing these with high-intensity sweeteners. As with all food products, the development of a formulation to replace sucrose is a complex proprietary process.

Use

Sugar substitutes are used instead of sugar for a number of reasons, including:

- To assist in weight loss – Some people choose to limit their food energy intake by replacing high-energy sugar or corn syrup with other sweeteners having little or no food energy. This allows them to eat the same foods they normally would while allowing them to lose weight and avoid other problems associated with excessive caloric intake.
- Dental care – Carbohydrates and sugars usually adhere to the tooth enamel, where bacteria feed upon them and quickly multiply. The bacteria convert the sugar to acids that decay the teeth. Sugar substitutes, unlike sugar, do not erode teeth as they are not fermented by the microflora of the dental plaque. A sweetener that can actually benefit dental health is xylitol, which tends to prevent bacteria from adhering to the tooth surface, thus preventing plaque formation and eventually decay. Xylitol cannot be fermented by bacteria that feed on sugar, so they have difficulty thriving, thus helping to prevent plaque formation.
- Diabetes mellitus – People with diabetes have difficulty regulating their blood sugar levels, and need to limit their sugar intake. Many artificial sweeteners allow sweet tasting food without increasing blood glucose. Others do release energy but are metabolized more slowly, preventing spikes in blood glucose.
- Reactive hypoglycemia – Individuals with reactive hypoglycemia will produce an excess of insulin after quickly absorbing glucose into the bloodstream. This causes their blood glucose levels to fall below the amount needed for proper body and brain function. As a result,

like diabetics, they must avoid intake of high-glycemic foods like white bread, and often use artificial sweeteners for sweetness without blood glucose.

- Cost and shelf life – Many sugar substitutes are cheaper than sugar. Sugar substitutes are often lower in total cost because of their long shelf-life and high sweetening intensity. This allows sugar substitutes to be used in products that will not perish after a short period of time.

Common Practice

The three primary compounds used as sugar substitutes in the United States are saccharin (e.g., *Sweet'N Low*), aspartame (e.g., *Equal, NutraSweet*) and sucralose (e.g., *Splenda, Altern*). Maltitol and sorbitol are often used, frequently in toothpaste, mouth wash, and in foods such as "no sugar added" ice cream. Erythritol is gaining momentum as a replacement for these other sugar alcohols in foods as it is much less likely to produce gastrointestinal distress when consumed in large amounts. In many other countries, xylitol, cyclamate, and the herbal sweetener stevia are used extensively.

When sweeteners are provided for restaurant customers to add to beverages such as tea and coffee themselves, they are often available in paper packets which can be torn and emptied. In North America, the colors are typically white for sucrose, blue for aspartame, pink for saccharin, yellow for sucralose (United States) or cyclamate (Canada), tan for turbinado, orange for monk fruit extract, and green for stevia.

Food Industry Use

The food and beverage industry is increasingly replacing sugar or corn syrup with artificial sweeteners in a range of products traditionally containing sugar.

According to market analysts Mintel, a total of 3,920 products containing artificial sweeteners were launched in the U.S. between 2000 and 2005. In 2004 alone, 1,649 artificially sweetened products were launched. According to market analysts Freedonia in 2012, the United States artificial sweetener market is set to grow at around 8% per year.

Aspartame is currently the most popular artificial sweetener in the U.S. food industry, as the price has dropped significantly since its patent registered by Monsanto Company expired in 1992. However, sucralose may soon replace it, as alternative processes to Tate & Lyle's patent seem to be emerging. According to Morgan Stanley, this can mean that the price of sucralose will drop by thirty percent.

Sugar substitutes are highly consumed in America. In 2003–2004, Americans two years of age and older consumed 585 grams (21 oz) per day of beverages and 375 grams (13 oz) per day of foods with caloric sweeteners. More than 66% of Americans consumed these beverages with sugar substitutes and 82.3% of Americans consumed foods with added caloric sweeteners. On the other hand, 10.8% of Americans in 2003–2004 consumed non-caloric sweetener flavored beverages and 5.8% consumed non-caloric sweetener flavored foods.

Some commonly consumed foods with sugar substitutes are diet sodas, cereals, and sugar-free desserts such as ice cream. Sugar substitutes are found in many products today due to their low or non-caloric characteristics. This can be used to market a product to dieters or those conscious of their sugar intake, such as consumers with diabetes. Sugar substitutes such as xylitol and saccharin have many positive research results that show qualities of dental decay prevention, which causes them to be popular for use in chewing gums and toothpaste.

Sugar Substitutes

Aspartame

Aspartame was discovered in 1965 by James M. Schlatter at the G.D. Searle company. He was working on an anti-ulcer drug and accidentally spilled some aspartame on his hand. When he licked his finger, he noticed that it had a sweet taste. It is an odorless, white crystalline powder that is derived from the two amino acids aspartic acid and phenylalanine. It is about 200 times as sweet as sugar and can be used as a tabletop sweetener or in frozen desserts, gelatins, beverages, and chewing gum. When cooked or stored at high temperatures, aspartame breaks down into its constituent amino acids. This makes aspartame undesirable as a baking sweetener. It is more stable in somewhat acidic conditions, such as in soft drinks. Though it does not have a bitter aftertaste like saccharin, it may not taste exactly like sugar. When eaten, aspartame is metabolized into its original amino acids. Because it is so intensely sweet, relatively little of it is needed to sweeten a food product, and is thus useful for reducing the number of calories in a product.

The safety of aspartame has been studied extensively since its discovery with research that includes animal studies, clinical and epidemiological research, and postmarketing surveillance, with aspartame being one of the most rigorously tested food ingredients to date. Aspartame has been subject to multiple claims against its safety, including supposed links to cancer as well as complaints of neurological or psychiatric side effects. Multiple peer-reviewed comprehensive review articles and independent reviews by governmental regulatory bodies have analyzed the published research on the safety of aspartame and have found aspartame is safe for consumption at current levels. Aspartame has been deemed safe for human consumption by over 100 regulatory agencies in their respective countries, including the UK Food Standards Agency, the European Food Safety Authority (EFSA) and Canada's Health Canada.

Cyclamate

In the United States, the Food and Drug Administration (FDA) banned the sale of cyclamate in 1969 after lab tests in rats involving a 10:1 mixture of cyclamate and saccharin indicated that large amounts of cyclamates causes bladder cancer, a disease to which rats are particularly susceptible. Cyclamates are still used as sweeteners in many parts of the world, including Europe (e.g. UK and Russia).

Cyclamate-based sugar substitute sold in Canada.

Saccharin

Aside from sugar of lead, saccharin was the first artificial sweetener and was originally synthesized in 1879 by Remsen and Fahlberg. Its sweet taste was discovered by accident. It had been created in an experiment with toluene derivatives. A process for the creation of saccharin from phthalic anhydride was developed in 1950, and, currently, saccharin is created by this process as well as the original process by which it was discovered. It is 300 to 500 times as sweet as sugar (sucrose) and is often used to improve the taste of toothpastes, dietary foods, and dietary beverages. The bitter aftertaste of saccharin is often minimized by blending it with other sweeteners.

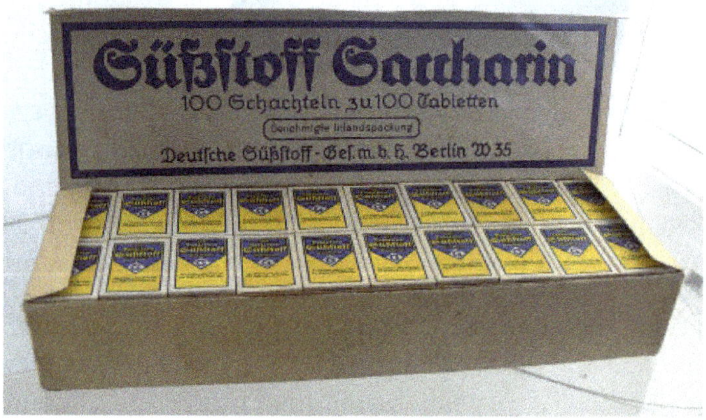

Saccharin, historical wrapping; Sugar Museum (Berlin)

Fear about saccharin increased when a 1960 study showed that high levels of saccharin may cause bladder cancer in laboratory rats. In 1977, Canada banned saccharin due to the animal research. In the United States, the FDA considered banning saccharin in 1977, but Congress stepped in and placed a moratorium on such a ban. The moratorium required a warning label and also mandated further study of saccharin safety.

Subsequent to this, it was discovered that saccharin causes cancer in male rats by a mechanism not found in humans. At high doses, saccharin causes a precipitate to form in rat urine. This precipitate damages the cells lining the bladder (urinary bladder urothelial cytotoxicity) and a tumor forms when the cells regenerate (regenerative hyperplasia). According to the International Agency for Research on Cancer, part of the World Health Organization, "Saccharin and its salts was [sic] downgraded from Group 2B, possibly carcinogenic to humans, to Group 3, not classifiable as to carcinogenicity to humans, despite sufficient evidence of carcinogenicity to animals, because it is carcinogenic by a non-DNA-reactive mechanism that is not relevant to humans because of critical interspecies differences in urine composition."

In 2001, the United States repealed the warning label requirement, while the threat of an FDA ban had already been lifted in 1991. Most other countries also permit saccharin, but restrict the levels of use, while other countries have outright banned it.

The EPA has officially removed saccharin and its salts from their list of hazardous constituents and commercial chemical products. In a 14 December 2010, release the EPA stated that saccharin is no longer considered a potential hazard to human health.

Stevia

Stevia has been widely used as a natural sweetener in South America for centuries and in Japan since 1970. Due to its unique characteristics of zero glycemic index and zero calories, it is fast becoming popular in many other countries. In 1987, the FDA issued a ban on stevia because it had not been approved as a food additive, although it continued to be available as a dietary supplement. After being provided with a significant amount of scientific data proving that there was no side-effect of using stevia as a sweetener from companies such as Cargill and Coca-Cola, the FDA gave a "no objection" approval for GRAS status in December 2008 to Truvia, a blend of rebiana and erythritol (developed by Cargill and The Coca-Cola Company), as well as PureVia (developed by PepsiCo and the Whole Earth Sweetener Company, a subsidiary of Merisant), both of which using rebaudioside A derived from the stevia plant. In Australia, the brand Vitarium have used Natvia, a natural stevia sweetener, to do a range on sugar-free children's milk mixes.

Sucralose

Sucralose is a chlorinated sugar that is about 600 times as sweet as sugar. It is produced from sucrose when three chlorine atoms replace three hydroxyl groups. It is used in beverages, frozen desserts, chewing gum, baked goods, and other foods. Unlike other artificial sweeteners, it is stable when heated and can therefore be used in baked and fried goods. The FDA approved sucralose in 1998.

Most of the controversy surrounding Splenda, a sucralose sweetener, is focused not on safety but on its marketing. It has been marketed with the slogan, "Splenda is made from sugar, so it tastes like sugar." Sucralose is prepared from either of two sugars, sucrose or raffinose. With either base sugar, processing replaces three oxygen-hydrogen groups in the sugar molecule with three chlorine atoms.

The "Truth About Splenda" website was created in 2005 by The Sugar Association, an organization representing sugar beet and sugar cane farmers in the United States, to provide its view of sucralose. In December 2004, five separate false-advertising claims were filed by the Sugar Association against Splenda manufacturers Merisant and McNeil Nutritionals for claims made about Splenda related to the slogan, "Made from sugar, so it tastes like sugar". French courts ordered the slogan to no longer be used in France, while in the U.S. the case came to an undisclosed settlement during the trial.

There are few safety concerns pertaining to sucralose and the way sucralose is metabolized suggests a reduced risk of toxicity. For example, sucralose is extremely insoluble in fat and, thus, does not accumulate in fatty tissues; sucralose also does not break down and will dechlorinate only under conditions that are not found during regular digestion (i.e., high heat applied to the powder form of the molecule). Only about 15% of sucralose is absorbed by the body and most of it passes out of the body unchanged.

Acesulfame Potassium

Acesulfame potassium (Ace-K) is 200 times sweeter than sucrose (common sugar), as sweet as aspartame, about two thirds as sweet as saccharin, and one third as sweet as sucralose. Like saccharin, it has a slightly bitter aftertaste, especially at high concentrations. Kraft Foods has patented the use of sodium ferulate to mask acesulfame's aftertaste. Acesulfame potassium is often blended with other sweeteners (usually aspartame or sucralose), which give a more sucrose-like taste, whereby each sweetener masks the other's aftertaste and also exhibits a synergistic effect in which the blend is sweeter than its components.

Unlike aspartame, acesulfame potassium is stable under heat, even under moderately acidic or basic conditions, allowing it to be used as a food additive in baking or in products that require a long shelf life. In carbonated drinks, it is almost always used in conjunction with another sweetener, such as aspartame or sucralose. It is also used as a sweetener in protein shakes and pharmaceutical products, especially chewable and liquid medications, where it can make the active ingredients more palatable.

Lead Acetate (Historic)

Lead acetate (sometimes called *sugar of lead*) is an artificial sugar substitute made from lead that is of historical interest because of its widespread use in the past, such as by ancient Romans. The use of lead acetate as a sweetener eventually produced lead poisoning in any individual ingesting it habitually. Lead acetate was abandoned as a food additive throughout most of the world after the high toxicity of lead compounds became apparent.

Mogrosides

More recently, mogrosides (typically extracted from monk fruit) have been used in commercial products after the FDA granted some of the compounds GRAS status in 2010. As of 2011, it is not (yet) a permitted sweetener in the EU, although it is allowed as a natural flavor at concentrations where it does not function as a sweetener. Some of the products incorporating it are Nestlé's Milo in Asia and certain Kellogg cereals in the U.S. It is also the basis of McNeil Nutritionals's tabletop sweetener Nectresse in the U.S. and Norbu Sweetener in Australia. As of 2012, the New Zealand company BioVittoria provides more than 90 percent of the global supply of monk fruit extract; its main manufacturing facility for the product is in Guilin, China.

Sweetness Relative to Sucrose

Natural Sugar Substitutes

The sweetnesses and energy densities are in comparison to those of sucrose.

Name	Sweetness by weight	Sweetness by food energy	Energy density	Notes
Brazzein	800			Protein
Curculin	550			Protein
Erythritol	0.7	14	0.05	
Glycyrrhizin	50			
Glycerol	0.6	0.55	1.075	E422

Name	Sweetness by weight	Sweetness by food energy	Energy density	Notes
Hydrogenated starch hydrolysates	0.4–0.9	0.5×–1.2	0.75	
Inulin				
Isomalt	0.45–0.65	0.9–1.3	0.5	E953
Lactitol	0.4	0.8	0.5	E966
Mogroside mix	300			
Mabinlin	100			Protein
Maltitol	0.9	1.7	0.525	E965
Malto-oligosaccharide				
Mannitol	0.5	1.2	0.4	E421
Miraculin				A protein that does not taste sweet by itself but modifies taste receptors to make sour things taste sweet temporarily
Monatin				Naturally occurring sweetener isolated from the plant *Sclerochiton ilicifolius*
Monellin	3,000			Protein; the sweetening ingredient in serendipity berries
Osladin				
Pentadin	500			Protein
Sorbitol	0.6	0.9	0.65	sugar alcohol, E420
Stevia	250			Extracts known as rebiana, Sweet and Fit Stevia, Truvia, PureVia; mainly containing rebaudioside A, a steviol glycoside
Tagatose	0.92	2.4	0.38	monosaccharide
Thaumatin	2,000			Protein; E957
Xylitol	1.0	1.7	0.6	E967

Artificial Sugar Substitutes

Artificial sweeteners contain little or no food energy, making a comparison of sweetness based on energy content not meaningful.

Name	Sweetness (by weight)	Trade name	Approval	Notes
Acesulfame potassium	200	Nutrinova	FDA 1988	E950 Hyet Sweet
Advantame	20,000		FDA	
Alitame	2,000		approved in Mexico, Australia, New Zealand and China.	Pfizer
Aspartame	160–200	NutraSweet, Equal	FDA 1981, EU-wide 1994	E951 Hyet Sweet
Salt of aspartame-acesulfame	350	Twinsweet		E962

Name	Sweetness (by weight)	Trade name	Approval	Notes
Sodium cyclamate	30		FDA Banned 1969, approved in EU and Canada	E952, Abbott
Dulcin	250		FDA Banned 1950	
Glucin	300			
Neohesperidin dihydrochalcone	1,500			E959
Neotame	8,000	NutraSweet	FDA 2002	E961
P-4000	4,000		FDA banned 1950	
Saccharin	300	Sweet'N Low	FDA 1958, Canada 2014	E954
Sucralose	600	Kaltame, Splenda	Canada 1991, FDA 1998, EU 2004	E955, Tate & Lyle

Health Effects

Insulin Response

A 2007 animal study indicated that a sweet taste induces an insulin response in rats. However, the extension of animal model findings to humans is unclear, as human studies of intragastric infusion of sucralose have shown no insulin response from analogous taste receptors.

A 2014 study by a collaboration of seventeen scientists from nine Israeli research institutes presented experimental evidence that artificial sweeteners may exacerbate, rather than prevent or mitigate, metabolic disorders such as Type 2 diabetes. They reported that artificial sweeteners increase the blood sugar levels in both mice and humans by altering the composition and function of the gut flora. Mice given drinking water supplemented with artificial sweetener (commercial formulations of saccharin, sucralose or aspartame) developed greater glucose intolerance than mice drinking pure water, or water with only sugar added. The effect occurred both in mice fed standard food and those on a high-fat diet. Changes in the composition of the gut flora were observed by sequencing a ribosomal RNA gene. When antibiotics were then used to kill off gut bacteria, the degree of glucose intolerance in mice fed either diet was restored to normal levels present before artificial sweetener was introduced. Human subjects were also studied; the gut bacteria from 381 non-diabetics averaging age 43 were analyzed, revealing differences in the gut bacteria between those subjects who habitually consumed artificial sweeteners and those who did not, as well as "markers" for diabetes, such as raised blood sugar levels and glucose intolerance. In a journal commentary, two researchers opined that artificial sweeteners "may contribute to, rather than alleviate, obesity-related metabolic conditions, by altering the composition and function of bacterial populations in the gut."

Cancer

As of 2015 it is unclear if artificial sweeteners affect the risk of cancer.

Comparison to Sugar

Eating natural sugars like glucose instead of a sugar substitute can also have negative health effects. The calories contained in sugar-sweetened beverages contributes to increases in body weight and body fat, and replacement of sugar by artificial sweeteners reduces weight. Obesity contributes to diabetes and cardiovascular disease. Glucose has a high glycemic index, sucrose medium, and fructose low. The consumption of added sugars has been positively associated with multiple measures known to increase cardiovascular disease risk amongst adolescents as well as adults. There is "convincing evidence from human intervention studies, epidemiological studies, animal studies and experimental studies, for an association between the amount and frequency of free sugars intake and dental caries" while other sugars (complex carbohydrate) consumption is normally associated with a lower rate of dental caries.

References

- Bucci, Luke (1995). Nutrition applied to injury rehabilitation and sports medicine. Boca Raton: CRC Press. p. 151. ISBN 0-8493-7913-X.
- Assessment of technologies for determining cancer risks from the environment. Darby, Pennsylvania, USA: DIANE publishing. 1981. p. 177. ISBN 1-4289-2437-X.
- König, J. (2015), "Food colour additives of synthetic origin", in Scotter, Michael J., Colour Additives for Foods and Beverages, Elsevier, pp. 35–60, doi:10.1016/B978-1-78242-011-8.00002-7, ISBN 978-1-78242-011-8
- Mitchell, Helen (2006). Sweeteners and sugar alternatives in food technology. Oxford, UK: Wiley-Blackwell. p. 94. ISBN 1-4051-3434-8
- "Recommendations on Wheat and Maize Flour Fortification Meeting Report: Interim Consensus Statement" (PDF). Who.int. Retrieved 2016-03-30.
- Arlt, Ulrike (29 Apr 2011). "The Legislation of Food Colours in Europe". The Natural Food Colours Association. Retrieved 18 Feb 2014.
- Downham, Alison; Collins, Paul (2000). "Colouring our foods in the last and next millennium" (PDF). International Journal of Food Science and Technology. Blackwell Science Ltd. 35: 5–22. doi:10.1046/j.1365-2621.2000.00373.x. Retrieved 18 Feb 2014.
- Hastings, Robert W. (January–March 1898). Hamilton, John B., ed. "Human Food Laws". Journal of the American Medical Association. Chicago: American Medical Association. 30 (1-13): 419–421. doi:10.1001/jama.1898.72440600019002e. Retrieved 17 Feb 2014.
- Meadows, Michelle (2006). "A Century of Ensuring Safe Foods and Cosmetics". FDA Consumer magazine. FDA (January–February). Retrieved 21 Feb 2014.
- Taylor Feehley; Cathryn R. Nagler (2014). "Artificial sweeteners induce glucose intolerance by altering the gut microbiota". Nature (preview). doi:10.1038/nature13752. Retrieved 17 September 2014.
- Tomaska LD and Brooke-Taylor, S. Food Additives - General pp 449-454 in Encyclopedia of Food Safety, Vol 2: Hazards and Diseases. Eds, Motarjemi Y et al. Academic Press, 2013. ISBN 9780123786135

4

Consequences of Food Contamination

Food contamination poses a serious threat as any mass produced food has a wide distribution radius. Food contamination can be caused due to faulty production techniques, improper storage or handling. This chapter deals exclusively with the topic of food contamination, foodborne illnesses and the various emerging foodborne pathogens. The chapter on food contamination offers an insightful focus, keeping in mind the complex subject matter.

Foodborne Illness

Foodborne illness (also foodborne disease and colloquially referred to as food poisoning) is any illness resulting from the food spoilage of contaminated food, pathogenic bacteria, viruses, or parasites that contaminate food, as well as chemical or natural toxins such as poisonous mushrooms and various species of beans that have not been boiled for at least 10 minutes.

Symptoms vary depending on the cause, and are described below in this article. A few broad generalizations can be made, e.g.: The incubation period ranges from hours to days, depending on the cause and on how much was consumed. The incubation period tends to cause sufferers to not associate the symptoms with the item consumed, and so to cause sufferers to attribute the symptoms to gastroenteritis for example. Symptoms often include vomiting, fever, and aches, and may include diarrhea. Bouts of vomiting can be repeated with an extended delay in between, because even if infected food was eliminated from the stomach in the first bout, microbes (if applicable) can pass through the stomach into the intestine via cells lining the intestinal walls and begin to multiply. Some types of microbes stay in the intestine, some produce a toxin that is absorbed into the bloodstream, and some can directly invade deeper body tissues.

Causes

Poorly stored food in a refrigerator

Foodborne illness usually arises from improper handling, preparation, or food storage. Good hygiene practices before, during, and after food preparation can reduce the chances of contracting an illness. There is a consensus in the public health community that regular hand-washing is one of the most effective defenses against the spread of foodborne illness. The action of monitoring food to ensure that it will not cause foodborne illness is known as food safety. Foodborne disease can also be caused by a large variety of toxins that affect the environment.

Foodborne illness can also be caused by pesticides or medicines in food and naturally toxic substances such as poisonous mushrooms or reef fish.

Bacteria

Bacteria are a common cause of foodborne illness. In the United Kingdom during 2000, the individual bacteria involved were the following: *Campylobacter jejuni* 77.3%, *Salmonella* 20.9%, *Escherichia coli* O157:H7 1.4%, and all others less than 0.56%. In the past, bacterial infections were thought to be more prevalent because few places had the capability to test for norovirus and no active surveillance was being done for this particular agent. Toxins from bacterial infections are delayed because the bacteria need time to multiply. As a result, symptoms associated with intoxication are usually not seen until 12–72 hours or more after eating contaminated food. However, in some cases, such as Staphylococcal food poisoning, the onset of illness can be as soon as 30 minutes after ingesting contaminated food.

Most common bacterial foodborne pathogens are:

- *Campylobacter jejuni* which can lead to secondary Guillain–Barré syndrome and periodontitis
- *Clostridium perfringens*, the "cafeteria germ"
- *Salmonella* spp. – its *S. typhimurium* infection is caused by consumption of eggs or poultry that are not adequately cooked or by other interactive human-animal pathogens
- *Escherichia coli O157:H7* enterohemorrhagic (EHEC) which can cause hemolytic-uremic syndrome

Other common bacterial foodborne pathogens are:

- *Bacillus cereus*
- *Escherichia coli*, other virulence properties, such as enteroinvasive (EIEC), enteropathogenic (EPEC), enterotoxigenic (ETEC), enteroaggregative (EAEC or EAgEC)
- *Listeria monocytogenes*
- *Shigella* spp.
- *Staphylococcus aureus*
- *Staphylococcal enteritis*
- *Streptococcus*

- *Vibrio cholerae*, including O1 and non-O1
- *Vibrio parahaemolyticus*
- *Vibrio vulnificus*
- *Yersinia enterocolitica* and *Yersinia pseudotuberculosis*

Less common bacterial agents:

- *Brucella* spp.
- *Corynebacterium ulcerans*
- *Coxiella burnetii* or Q fever
- *Plesiomonas shigelloides*

Enterotoxins

In addition to disease caused by direct bacterial infection, some foodborne illnesses are caused by enterotoxins (exotoxins targeting the intestines). Enterotoxins can produce illness even when the microbes that produced them have been killed. Symptom appearance varies with the toxin but may be rapid in onset, as in the case of enterotoxins of *Staphylococcus aureus* in which symptoms appear in one to six hours. This causes intense vomiting including or not including diarrhea (resulting in staphylococcal enteritis), and staphylococcal enterotoxins (most commonly staphylococcal enterotoxin A but also including staphylococcal enterotoxin B) are the most commonly reported enterotoxins although cases of poisoning are likely underestimated. It occurs mainly in cooked and processed foods due to competition with other biota in raw foods, and humans are the main cause of contamination as a substantial percentage of humans are persistent carriers of *S. aureus*. The CDC has estimated about 240,000 cases per year in the United States.

- *Clostridium botulinum*
- *Clostridium perfringens*
- *Bacillus cereus*

The rare but potentially deadly disease botulism occurs when the anaerobic bacterium *Clostridium botulinum* grows in improperly canned low-acid foods and produces botulin, a powerful paralytic toxin.

Pseudoalteromonas tetraodonis, certain species of Pseudomonas and Vibrio, and some other bacteria, produce the lethal tetrodotoxin, which is present in the tissues of some living animal species rather than being a product of decomposition.

Emerging Foodborne Pathogens

Many foodborne illnesses remain poorly understood.

- *Aeromonas hydrophila, Aeromonas caviae, Aeromonas sobria*

Preventing Bacterial Food Poisoning

Prevention is mainly the role of the state, through the definition of strict rules of hygiene and a public services of veterinary surveying of animal products in the food chain, from farming to the transformation industry and delivery (shops and restaurants). This regulation includes:

Proper storage and refrigeration of food help in the prevention of food poisoning

- traceability: in a final product, it must be possible to know the origin of the ingredients (originating farm, identification of the harvesting or of the animal) and where and when it was processed; the origin of the illness can thus be tracked and solved (and possibly penalized), and the final products can be removed from the sale if a problem is detected;
- enforcement of hygiene procedures such as HACCP and the "cold chain";
- power of control and of law enforcement of veterinarians.

In August 2006, the United States Food and Drug Administration approved Phage therapy which involves spraying meat with viruses that infect bacteria, and thus preventing infection. This has raised concerns, because without mandatory labelling consumers would not be aware that meat and poultry products have been treated with the spray.

At home, prevention mainly consists of good food safety practices. Many forms of bacterial poisoning can be prevented by cooking it sufficiently, and either eating it quickly or refrigerating it effectively. Many toxins, however, are not destroyed by heat treatment.

Techniques that help prevent food borne illness in the kitchen are hand washing, preventing cross-contamination, proper storage, and maintaining cooking temperatures. In general, freezing or refrigerating prevents virtually all bacteria from growing, and heating food sufficiently kills parasites, viruses, and most bacteria. Bacteria grow most rapidly at the range of temperatures between 40 °F and 140 °F, called the "danger zone". Storing food below or above the "danger zone" can effectively limit the production of toxins. For storing leftovers, the food must be put in shallow containers for quick cooling and must be refrigerated within two hours. When food is reheated, it must reach an internal temperature of 165 °F (74 °C) or until hot or steaming to kill bacteria.

Mycotoxins and Alimentary Mycotoxicoses

The term alimentary mycotoxicoses refers to the effect of poisoning by Mycotoxins (The term 'mycotoxin' is usually reserved for the toxic chemical products produced by fungi that readily colonize crops) through food consumption. Mycotoxins sometimes have important effects on human and animal health. For example, an outbreak which occurred in the UK in 1960 caused the death of 100,000 turkeys which had consumed aflatoxin-contaminated peanut meal. In the USSR in World War II, 5,000 people died due to Alimentary Toxic Aleukia (ALA). The common foodborne Mycotoxins include:

- Aflatoxins – originated from Aspergillus parasiticus and Aspergillus flavus. They are frequently found in tree nuts, peanuts, maize, sorghum and other oilseeds, including corn and cottonseeds. The pronounced forms of Aflatoxins are those of B1, B2, G1, and G2, amongst which Aflatoxin B1 predominantly targets the liver, which will result in necrosis, cirrhosis, and carcinoma. In the US, the acceptable level of total aflatoxins in foods is less than 20 μg/kg, except for Aflatoxin M1 in milk, which should be less than 0.5 μg/kg. The official document can be found at FDA's website.
- Altertoxins – are those of Alternariol (AOH), Alternariol methyl ether (AME), Altenuene (ALT), Altertoxin-1 (ATX-1), Tenuazonic acid (TeA) and Radicinin (RAD), originated from Alternaria spp. Some of the toxins can be present in sorghum, ragi, wheat and tomatoes. Some research has shown that the toxins can be easily cross-contaminated between grain commodities, suggesting that manufacturing and storage of grain commodities is a critical practice.
- Citrinin
- Citreoviridin
- Cyclopiazonic acid
- Cytochalasins
- Ergot alkaloids / Ergopeptine alkaloids – Ergotamine
- Fumonisins – Crop corn can be easily contaminated by the fungi Fusarium moniliforme, and its Fumonisin B1 will cause Leukoencephalomalacia (LEM) in horses, Pulmonary edema syndrome (PES) in pigs, liver cancer in rats and Esophageal cancer in humans. For human and animal health, both the FDA and the EC have regulated the content levels of toxins in food and animal feed.
- Fusaric acid
- Fusarochromanone
- Kojic acid
- Lolitrem alkaloids
- Moniliformin
- 3-Nitropropionic acid

- Nivalenol
- Ochratoxins – In Australia, The Limit of Reporting (LOR) level for Ochratoxin A (OTA) analyses in 20th Australian Total Diet Survey was 1 μg/kg, whereas the EC restricts the content of OTA to 5 μg/kg in cereal commodities, 3 μg/kg in processed products and 10 μg/kg in dried vine fruits.
- Oosporeine
- Patulin – Currently, this toxin has been advisably regulated on fruit products. The EC and the FDA have limited it to under 50 μg/kg for fruit juice and fruit nectar, while limits of 25 μg/kg for solid-contained fruit products and 10 μg/kg for baby foods were specified by the EC.
- Phomopsins
- Sporidesmin A
- Sterigmatocystin
- Tremorgenic mycotoxins – Five of them have been reported to be associated with molds found in fermented meats. These are Fumitremorgen B, Paxilline, Penitrem A, Verrucosidin, and Verruculogen.
- Trichothecenes – sourced from Cephalosporium, Fusarium, Myrothecium, Stachybotrys and Trichoderma. The toxins are usually found in molded maize, wheat, corn, peanuts and rice, or animal feed of hay and straw. Four trichothecenes, T-2 toxin, HT-2 toxin, diacetoxyscirpenol (DAS) and deoxynivalenol (DON) have been most commonly encountered by humans and animals. The consequences of oral intake of, or dermal exposure to, the toxins will result in Alimentary toxic aleukia, neutropenia, aplastic anemia, thrombocytopenia and/or skin irritation. In 1993, the FDA issued a document for the content limits of DON in food and animal feed at an advisory level. In 2003, US published a patent that is very promising for farmers to produce a trichothecene-resistant crop.
- Zearalenone
- Zearalenols

Viruses

Viral infections make up perhaps one third of cases of food poisoning in developed countries. In the US, more than 50% of cases are viral and noroviruses are the most common foodborne illness, causing 57% of outbreaks in 2004. Foodborne viral infection are usually of intermediate (1–3 days) incubation period, causing illnesses which are self-limited in otherwise healthy individuals; they are similar to the bacterial forms described above.

- Enterovirus
- Hepatitis A is distinguished from other viral causes by its prolonged (2–6 week) incubation period and its ability to spread beyond the stomach and intestines into the liver. It often results in jaundice, or yellowing of the skin, but rarely leads to chronic liver dysfunction.

The virus has been found to cause infection due to the consumption of fresh-cut produce which has fecal contamination.

- Hepatitis E
- Norovirus
- Rotavirus

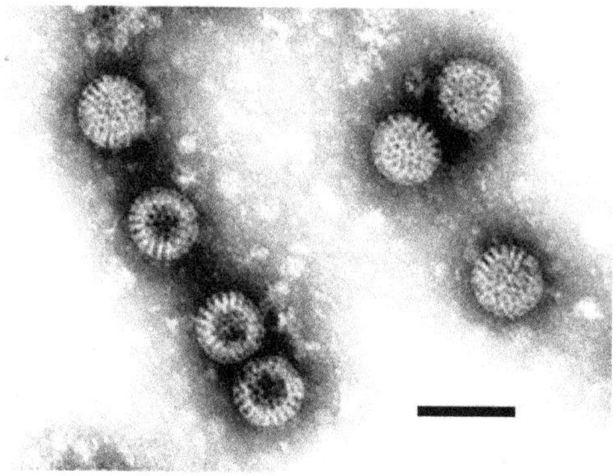

Rotavirus

Parasites

Most foodborne parasites are zoonoses.

- Platyhelminthes:
    - *Diphyllobothrium* sp.
    - *Nanophyetus* sp.
    - *Taenia saginata*
    - *Taenia solium*
    - *Fasciola hepatica*
- Nematode:
    - *Anisakis* sp.
    - *Ascaris lumbricoides*
    - *Eustrongylides* sp.
    - *Trichinella spiralis*
    - *Trichuris trichiura*
- Protozoa:
    - *Acanthamoeba* and other free-living amoebae

- *Cryptosporidium parvum*
- *Cyclospora cayetanensis*
- *Entamoeba histolytica*
- *Giardia lamblia*
- *Sarcocystis hominis*
- *Sarcocystis suihominis*
- *Toxoplasma gondii*

Natural Toxins

Several foods can naturally contain toxins, many of which are not produced by bacteria. Plants in particular may be toxic; animals which are naturally poisonous to eat are rare. In evolutionary terms, animals can escape being eaten by fleeing; plants can use only passive defenses such as poisons and distasteful substances, for example capsaicin in chili peppers and pungent sulfur compounds in garlic and onions. Most animal poisons are not synthesised by the animal, but acquired by eating poisonous plants to which the animal is immune, or by bacterial action.

- Alkaloids
- Ciguatera poisoning
- Grayanotoxin (honey intoxication)
- Mushroom toxins
- Phytohaemagglutinin (red kidney bean poisoning; destroyed by boiling)
- Pyrrolizidine alkaloids
- Shellfish toxin, including paralytic shellfish poisoning, diarrhetic shellfish poisoning, neurotoxic shellfish poisoning, amnesic shellfish poisoning and ciguatera fish poisoning
- Scombrotoxin
- Tetrodotoxin (fugu fish poisoning)

Some plants contain substances which are toxic in large doses, but have therapeutic properties in appropriate dosages.

- Foxglove contains cardiac glycosides.
- Poisonous hemlock (conium) has medicinal uses.

Other Pathogenic Agents

- Prions, resulting in Creutzfeldt–Jakob disease

"Ptomaine Poisoning"

In 1883, the Italian professor Salmi, of Bologna, introduced the generic name *ptomaine* (from Greek *ptōma*, "fall, fallen body, corpse") for alkaloids found in decaying animal and vegetable matter, especially (as reflected in their names) putrescine and cadaverine. The 1892 *Merck's Bulletin* stated, "We name such products of bacterial origin ptomaines; and the special alkaloid produced by the comma bacillus is variously named Cadaverine, Putrescine, etc." While *The Lancet* stated, "The chemical ferments produced in the system, the...ptomaines which may exercise so disastrous an influence." It is now known that the "disastrous...influence" is due to the direct action of bacteria and only slightly to the alkaloids. Thus, the use of the phrase "ptomaine poisoning" is now obsolete.

Tainted potato salad sickening hundreds at a Communist political convention in Massillon, Ohio, and aboard a Washington DC cruise boat in separate incidents during a single week in 1932 drew national attention to the dangers of so-called "ptomaine poisoning" in the pages of the American news weekly, *Time*. Another newspaper article from 1944 told of more than 150 persons being hospitalized in Chicago with ptomaine poisoning apparently from rice pudding served by a chain of restaurants.

Mechanism

Incubation Period

The delay between consumption of a contaminated food and appearance of the first symptoms of illness is called the incubation period. This ranges from hours to days (and rarely months or even years, such as in the case of listeriosis or bovine spongiform encephalopathy), depending on the agent, and on how much was consumed. If symptoms occur within one to six hours after eating the food, it suggests that it is caused by a bacterial toxin or a chemical rather than live bacteria.

The long incubation period of many foodborne illnesses tends to cause sufferers to attribute their symptoms to gastroenteritis.

During the incubation period, microbes pass through the stomach into the intestine, attach to the cells lining the intestinal walls, and begin to multiply there. Some types of microbes stay in the intestine, some produce a toxin that is absorbed into the bloodstream, and some can directly invade the deeper body tissues. The symptoms produced depend on the type of microbe.

Infectious Dose

The infectious dose is the amount of agent that must be consumed to give rise to symptoms of foodborne illness, and varies according to the agent and the consumer's age and overall health. In the case of *Salmonella* a relatively large inoculum of 1 million to 1 billion organisms is necessary to produce symptoms in healthy human volunteers, as *Salmonellae* are very sensitive to acid. An unusually high stomach pH level (low acidity) greatly reduces the number of bacteria required to cause symptoms by a factor of between 10 and 100.

Epidemiology

United States

In the United States, using FoodNet data from 2000–2007, the CDCP estimated there were 47.8 million foodborne illnesses per year (16,000 cases for 100,000 inhabitants) with 9.4 million of these caused by 31 known identified pathogens.

- 127,839 were hospitalized (43 per 100,000 inhabitants).
- 3,037 people died (1.0 per 100,000 inhabitants).

Causes of foodborne illness in U. S.				Causes of death by foodborne illness in U. S.			
	Cause	**Annual cases**	**rate (per 100,000 inhabitants)**		**Cause**	**Annual deaths**	**Rate (per 100,000 inhabitants)**
1	Norovirus	5,461,731 cases	X	1	Salmonella	378 cases	0.126
2	Salmonella	1,027,561 cases	X	2	Toxoplasma gondii	327 cases	0.109
3	Clostridium perfringens	965,958 cases	X	3	Listeria	255 cases	0.085
4	Campylobacter	845,024 cases	X	4	Norovirus	149 cases	0.050

France

This data pertains to reported medical cases of 23 specific pathogens, as opposed to total population estimates of all food-borne illness for the United States.

In France, for 750,000 cases (1210 per 100,000 inhabitants):

- 70,000 people consulted in the emergency department of an hospital (113 per 100,000 inhabitants.);
- 113,000 people were hospitalized (182 per 100,000 inhabitants);
- 460 people died (0.75 per 100,000 inhabitants).

Australia

A study by the Australian National University, published in November 2014, found in 2010 that there were an estimated 4.1 million cases of foodborne gastroenteritis acquired in Australia on average each year, along with 5,140 cases of non-gastrointestinal illness. The study was funded by the Australian Department of Health, Food Standards Australia New Zealand and the NSW Food Authority.

The main causes were Norovirus, pathogenic Escherichia coli, Campylobacter spp. and non-typhoidal Salmonella spp., although the causes of approximately 80% of illnesses were unknown. Approximately 25% (90% CrI: 13%–42%) of the 15.9 million episodes of gastroenteritis that occur in Australia were estimated to be transmitted by contaminated food. This equates to an average of approximately one episode of foodborne gastroenteritis every five years per person. Data on the number of hospitalisations and deaths represent the occurrence of serious foodborne illness. Including gastroenteritis, non-gastroenteritis and sequelae, there were an estimated annual 31,920 (90% CrI: 29,500–35,500) hospitalisations due to foodborne illness and 86 (90% CrI: 70–105) deaths due to foodborne illness circa 2010. This study concludes that these rates are similar to recent estimates in the US and Canada.

Causes of foodborne illness in France				Causes of death by foodborne illness in France			
	Cause	**Annual hospitalizations**	**Rate (per 100,000 inhabitants)**		**Cause**	**Annual deaths**	**Rate (per 100,000 inhabitants)**
1	Salmonella	8,000 cases	13	1	Salmonella	~300 cases	0.5
2	Campylobacter	3,000 cases	4.8	2	Listeria	~80 cases	0.13
3	Parasites incl. Toxoplasma	500 cases 400 cases	0.8 0.65	3	Parasites	~37 cases	0.06 (95% due to toxoplasma)
4	Listeria	300 cases	0.5	4	Campylobacter	~15 cases	0.02
5	Hepatitis A	60 cases	0.1	5	Hepatitis A	~2 cases	0.003

A main aim of this study was to compare if foodborne illness incidence had increased over time. In this study, similar methods of assessment were applied to data from circa 2000, which showed that the rate of foodborne gastroenteritis had not changed significantly over time. Two key estimates were the total number of gastroenteritis episodes each year, and the proportion considered foodborne. In circa 2010, it was estimated that 25% of all episodes of gastroenteritis were foodborne. By applying this proportion of episodes due to food to the incidence of gastroenteritis circa 2000, there were an estimated 4.3 million (90% CrI: 2.2–7.3 million) episodes of foodborne gastroenteritis circa 2000, although credible intervals overlap with 2010. Taking into account changes in population size, applying these equivalent methods suggests a 17% decrease in the rate of foodborne gastroenteritis between 2000 and 2010, with considerable overlap of the 90% credible intervals.

This study replaces a previous estimate of 5.4 million cases of food-borne illness in Australia every year, causing:

- 18,000 hospitalizations
- 120 deaths (0.5 deaths per 100,000 inhabitants)
- 2.1 million lost days off work

- 1.2 million doctor consultations
- 300,000 prescriptions for antibiotics.

Most foodborne disease outbreaks in Australia have been linked to raw or minimally cooked eggs or poultry. The Australian Food Safety Information Council estimates that one third of cases of food poisoning occur in the home

Comparison between countries

Country	Annual deaths per 100,000 inhabitants	Annual hospitalization per 100,000 inhabitants
USA	1.0	43
France	0.75	182
Australia	0.5	82

Outbreaks

The vast majority of reported cases of foodborne illness occur as individual or sporadic cases. The origin of most sporadic cases is undetermined. In the United States, where people eat outside the home frequently, 58% of cases originate from commercial food facilities (2004 FoodNet data). An outbreak is defined as occurring when two or more people experience similar illness after consuming food from a common source.

Often, a combination of events contributes to an outbreak, for example, food might be left at room temperature for many hours, allowing bacteria to multiply which is compounded by inadequate cooking which results in a failure to kill the dangerously elevated bacterial levels.

Outbreaks are usually identified when those affected know each other. However, more and more, outbreaks are identified by public health staff from unexpected increases in laboratory results for certain strains of bacteria. Outbreak detection and investigation in the United States is primarily handled by local health jurisdictions and is inconsistent from district to district. It is estimated that 1–2% of outbreaks are detected.

Society and Culture

United Kingdom

In postwar Aberdeen (1964) a large-scale (>400 cases) outbreak of typhoid occurred, caused by contaminated corned beef which had been imported from Argentina. The corned beef was placed in cans and because the cooling plant had failed, cold river water from the Plate estuary was used to cool the cans. One of the cans had a defect and the meat inside was contaminated. This meat was then sliced using a meat slicer in a shop in Aberdeen, and a lack of cleaning the machinery led to spreading the contamination to other meats cut in the slicer. These meats were then eaten by the people of Aberdeen who then became ill.

Serious outbreaks of foodborne illness since the 1970s prompted key changes in UK food safety law. These included the death of 19 patients in the Stanley Royd Hospital outbreak and the bovine spongiform encephalopathy (BSE, mad cow disease) outbreak identified in the 1980s. The death of 17 people in the 1996 Wishaw outbreak of *E. coli* O157 was a precursor to the establishment of the Food Standards Agency which, according to Tony Blair in the 1998 white paper *A Force for Change* Cm 3830, "would be powerful, open and dedicated to the interests of consumers".

In May 2015, for the second year running, England's Food Standards Agency devoted its annual Food Safety Week to – "The Chicken Challenge". The focus was on the handling of raw chicken in the home and in catering facilities in a drive to reduce the worryingly high levels of food poisoning from the *campylobacter* bacterium. Anne Hardy argues that widespread public education of food hygiene can be useful, particularly through media (T.V cookery programmes) and advertisement. She points to the examples set by Scandinavian societies.

United States

In 2001, the Center for Science in the Public Interest petitioned the United States Department of Agriculture to require meat packers to remove spinal cords before processing cattle carcasses for human consumption, a measure designed to lessen the risk of infection by variant Creutzfeldt-Jakob disease. The petition was supported by the American Public Health Association, the Consumer Federation of America, the Government Accountability Project, the National Consumers League, and Safe Tables Our Priority. This was opposed by the National Cattlemen's Beef Association, the National Renderers Association, the National Meat Association, the Pork Producers Council, sheep raisers, milk producers, the Turkey Federation, and eight other organizations from the animal-derived food industry.

None of the US Department of Health and Human Services targets regarding incidence of foodborne infections were reached in 2007.

Organizations

The World Health Organization Department of Food Safety and Zoonoses (FOS) provides scientific advice for organizations and the public on issues concerning the safety of food. Its mission is to lower the burden of foodborne disease, thereby strengthening the health security and sustainable development of Member States. Foodborne and waterborne diarrhoeal diseases kill an estimated 2.2 million people annually, most of whom are children. WHO works closely with the Food and Agriculture Organization of the United Nations (FAO) to address food safety issues along the entire food production chain—from production to consumption—using new methods of risk analysis. These methods provide efficient, science-based tools to improve food safety, thereby benefiting both public health and economic development.

International Food Safety Authorities Network (INFOSAN)

The International Food Safety Authorities Network (INFOSAN) is a joint program of the WHO and FAO. INFOSAN has been connecting national authorities from around the globe since 2004, with the goal of preventing the international spread of contaminated food and foodborne disease and strengthening food safety systems globally. This is done by:

1. Promoting the rapid exchange of information during food safety events;
2. Sharing information on important food safety issues of global interest;

3. Promoting partnership and collaboration between countries; and
4. Helping countries strengthen their capacity to manage food safety risks.

Membership to INFOSAN is voluntary, but is restricted to representatives from national and regional government authorities and requires an official letter of designation. INFOSAN seeks to reflect the multidisciplinary nature of food safety and promote intersectoral collaboration by requesting the designation of Focal Points in each of the respective national authorities with a stake in food safety, and a single Emergency Contact Point in the national authority with the responsibility for coordinating national food safety emergencies; countries choosing to be members of INFOSAN are committed to sharing information between their respective food safety authorities and other INFOSAN members. The operational definition of a food safety authority includes those authorities involved in: food policy; risk assessment; food control and management; food inspection services; foodborne disease surveillance and response; laboratory services for monitoring and surveillance of foods and foodborne diseases; and food safety information, education and communication across the farm-to-table continuum.

Regulatory Steps

Food may be contaminated during all stages of food production and retailing. In order to prevent viral contamination, regulatory authorities in Europe have enacted several measures:

- European Commission Regulation (EC) No 2073/2005 of November 15, 2005
- European Committee for Standardization (CEN): Standard method for the detection of norovirus and hepatitis A virus in food products (shellfish, fruits and vegetables, surfaces and bottled water)
- CODEX Committee on Food Hygiene (CCFH): Guideline for the application of general principles of food hygiene for the control of viruses in food

Various Emerging Foodborne Pathogens

Aeromonas Hydrophila

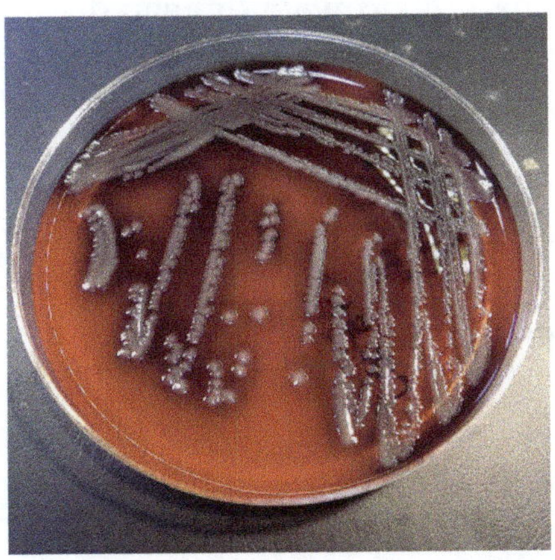

Aeromonas hydrophila colonies on the blood agar.

Aeromonas hydrophila is a heterotrophic, Gram-negative, rod-shaped bacterium mainly found in areas with a warm climate. This bacterium can be found in fresh or brackish water. It can survive in aerobic and anaerobic environments, and can digest materials such as gelatin and hemoglobin. *A. hydrophila* was isolated from humans and animals in the 1950s. It is the most well known of the species of *Aeromonas*. It is resistant to most common antibiotics and cold temperatures and is oxidase and indole positive.

Structure

A. hydrophila bacteria are Gram-negative, straight rods with rounded ends (bacilli to coccibacilli shape) usually from 0.3 to 1.0 μm in width, and 1.0 to 3.0 μm in length. They can grow at temperatures as low as 4 °C. These bacteria are motile by a polar flagellum.

Pathology

Because of its structure, it is very toxic to many organisms. When it enters the body of its victim, it travels through the bloodstream to the first available organ. It produces aerolysin cytotoxic enterotoxin that can cause tissue damage. *A. hydrophila, A. caviae,* and *A. sobria* are all considered to be opportunistic pathogens, meaning they rarely infect healthy individuals. *A. hydrophila* is widely considered a major fish and amphibian pathogen, and its pathogenicity in humans has been recognized for decades. The genomic insights of aeromonads could be a stepping stone into understanding of them

Pathogenic Mechanism

The pathogenicity of *Aeromonas* species was believed to be mediated by a number of extracellular proteins such as aerolysin, lipase, chitinase, amylase, gelatinase, hemolysins, and enterotoxins. However, the pathogenic mechanisms are unknown. The recently proposed type-III secretion system (TTSS) has been linked to *Aeromonas* pathogenesis. TTSS is a specialized protein secretion machinery that exports virulence factors directly to host cells. These factors subvert normal host cell functions to the benefit of invading bacteria. In contrast to the general secretory pathway, the TTSS is triggered when a pathogen comes in contact with host cells. ADP-ribosylation toxin is one of the effector molecules secreted by several pathogenic bacteria and translocated through the TTSS and delivered into the host cytoplasm, which leads to interruption of the NF-κB pathway, cytoskeletal damage, and apoptosis. This toxin has been characterized in *A. hydrophila* (human diarrhoeal isolate), *A. salmonicida* (fish pathogen), and *A. jandaei* GV17, a pathogenic strain which can cause disease both in humans and fish.

Occurrence of Exposure

A. hydrophila infections occur most often during environmental changes, stressors, changes in temperature, in contaminated environments, and when an organism is already infected with a virus or another bacterium. It can also be ingested through food products contaminated with the bacterium, such as seafood, meats, and even certain vegetables such as sprouts. It can also be transmitted by leeches.

Fish and Amphibians

A. hydrophila is associated with diseases mainly found in freshwater fish and amphibians, because these organisms live in aquatic environments. It is linked to a disease found in frogs called red leg, which causes internal, sometimes fatal hemorrhaging. When infected with *A. hydrophila*, fish develop ulcers, tail rot, fin rot, and hemorrhagic septicemia. Hemorrhagic septicaemia causes lesions that lead to scale shedding, hemorrhages in the gills and anal area, ulcers, exophthalmia, and abdominal swelling.

Human Diseases

A. hydrophila is not as pathogenic to humans as it is to fish and amphibians. One of the diseases it can cause in humans, gastroenteritis, occurs mostly in young children and people who have compromised immune systems or growth problems. This bacterium is linked to two types of gastroenteritis. The first type is a disease similar to cholera, which causes rice-water diarrhea. The other type is dysenteric gastroenteritis, which causes loose stools filled with blood and mucus. Dysenteric gastroenteritis is the most severe out of the two types, and can last for several weeks. *A. hydrophila* is also associated with cellulitis. It also causes diseases such as myonecrosis and eczema in people with compromised or suppressed (by medication) immune systems. In very rare cases, *A. hydrophila* can cause necrotizing fasciitis.

Outbreaks

Though *A. hydrophila* can cause serious disease, large scale outbreaks have not been reported. Outbreaks among vertebrates have occurred. One such incident occurred in Puerto Rico inside the intestinal tracts of lizards. Some 116 different strains were found in the lizards. On May 1, 1988, a small outbreak happened in California. The 225 isolates in 219 patients caused their hospital admissions. Confidential morbidity report cards were used to report the cases to the local health departments. Investigations were conducted, and reports were sent to the California Department of Health Services for diagnosis and methods in treatment.

Treatments

A. hydrophila can be eliminated using a 1% sodium hypochlorite solution or 2% calcium hypochlorite solution.

Brage et al., 1990 recommends fluoroquinolone administration as a prophylactic treatment during medicinal leech application.

Antibiotics such as chloramphenicol, florenicol, tetracycline, sulfonamide, nitrofuran derivatives, and pyrodinecarboxylic acids are used to eliminate and control the infection of *A. hydrophila*.

Terramycin is placed in fish food during hatchery operations as another chemotherapeutic agent in preventing *A. hydrophila*.

Zoonosis

Zoonoses are infectious diseases of animals (usually vertebrates) that can naturally be transmitted to humans.

Major modern diseases such as Ebola virus disease, salmonellosis and influenza are zoonoses. Zoonoses can be caused by a range of disease pathogens such as viruses, bacteria, fungi and parasites;

of 1,415 pathogens known to infect humans, 61% were zoonotic. Most human diseases originated in animals; however, only diseases that routinely involve animal to human transmission, like rabies, are considered as zoonoses.

Zoonoses have different modes of transmission. In direct zoonosis the disease is directly transmitted from animals to humans through media such as air (influenza) or through bites and saliva (rabies). In contrast, transmission can also occur via an intermediate species (referred to as a vector), which carry the disease pathogen without getting infected. When humans infect animals, it is called reverse zoonosis or anthroponosis.

History

During most of human prehistory groups of hunter-gatherers were probably very small. Such groups probably made contact with other such bands only rarely. Such isolation would have caused epidemic diseases to be restricted to any given local population, because propagation and expansion of epidemics depend on frequent contact with other individuals who have not yet developed an adequate immune response. To persist in such a population, a pathogen either had to be a chronic infection, staying present and potentially infectious in the infected host for long periods, or it had to have other additional species as reservoir where it can maintain itself until further susceptible hosts are contacted and infected. In fact, for many 'human' diseases, the human is actually better viewed as an accidental or incidental victim and a dead-end host. Examples include rabies, anthrax, tularemia and West Nile virus. Thus, much of human exposure to infectious disease has been zoonotic.

Many modern diseases, even epidemic diseases, started out as zoonotic diseases. It is hard to establish with certainty which diseases jumped from other animals to humans, but there is increasing evidence from DNA and RNA sequencing, that measles, smallpox, influenza, HIV, and diphtheria came to us this way. Various forms of the common cold and tuberculosis also are adaptations of strains originating in other species.

Zoonoses are of interest because they are often previously unrecognized diseases or have increased virulence in populations lacking immunity. The West Nile virus appeared in the United States in 1999 in the New York City area, and moved through the country in the summer of 2002, causing much distress. Bubonic plague is a zoonotic disease, as are salmonellosis, Rocky Mountain spotted fever, and Lyme disease.

A major factor contributing to the appearance of new zoonotic pathogens in human populations is increased contact between humans and wildlife. This can be caused either by encroachment of human activity into wilderness areas or by movement of wild animals into areas of human activity. An example of this is the outbreak of Nipah virus in peninsular Malaysia in 1999, when intensive pig farming began on the habitat of infected fruit bats. Unidentified infection of the pigs amplified the force of infection, eventually transmitting the virus to farmers and causing 105 human deaths.

Similarly, in recent times avian influenza and West Nile virus have spilled over into human populations probably due to interactions between the carrier host and domestic animals. Highly mobile animals such as bats and birds may present a greater risk of zoonotic transmission than other animals due to the ease with which they can move into areas of human habitation.

Because they depend on the human host for part of their life-cycle, diseases such as African schistosomiasis, river blindness, and elephantiasis are *not* defined as zoonotic, even though they may depend on transmission by insects or other vectors.

Foodborne Illness

The most significant zoonotic pathogens causing foodborne diseases are *Escherichia coli* O157:H7, *Campylobacter*, *Caliciviridae*, and *Salmonella*.

In 2006, a conference held in Berlin was focusing on the issue of zoonotic pathogen effects on food safety, urging governments to intervene, and the public to be vigilant towards the risks of catching food-borne diseases from farm-to-dining table.

Many food outbreaks can be linked to zoonotic pathogens. Many different types of food can be contaminated that have an animal origin. Some common foods linked to zoonotic contaminations include eggs, seafood, meat, dairy, and even some vegetables. Food outbreaks should be handled in preparedness plans to prevent widespread outbreaks and to efficiently and effectively contain outbreaks.

Lists of Diseases

Disease	Pathogen(s)	Animals involved	Mode of transmission
African sleeping sickness	*Trypanosoma brucei rhodesiense*	range of wild animals and domestic livestock	transmitted by the bite of the tsetse fly
Anthrax	*Bacillus anthracis*	commonly - grazing herbivores such as cattle, sheep, goats, camels, horses, and pigs	by ingestion, inhalation or skin contact of spores
Brucellosis	*Brucella* spp.	cattle, goats	infected milk or meat
Cat-scratch disease	*Bartonella henselae, Bartonella quintana*	cats	bites or scratches from infected cats
Variant Creutzfeldt–Jakob disease	PrPvCJD	cattle	eating meat from animals with bovine spongiform encephalopathy (BSE)
Cysticercosis & Taeniasis	*Taenia solium, Taenia saginata*	commonly - pigs and cattle	consuming water or food contaminated with the tapeworm eggs (cysticercosis) or raw or undercooked pork contaminated with the cysticerci (taeniasis)
Cryptococcosis	*Cryptococcus neoformans*	commonly - birds like pigeons	inhaling fungi
Ebola virus disease (a haemorrhagic fever)	*Ebolavirus* spp.	chimpanzees, gorillas, fruit bats, monkeys, forest antelope and porcupines	through body fluids, organs

Other haemorrhagic fevers (Marburg viral haemorrhagic fever, Lassa fever, Crimean-Congo haemorrhagic fever, Rift Valley fever)	Varies - commonly viruses	varies (sometimes unknown) - commonly camels, hares, hedgehogs, cattle, sheep, goats, horses and swine	infection usually occurs through direct contact with infected animals
Echinococcosis	*Echinococcus* spp.	commonly - dogs, foxes, wolves, sheep, and rodents	eating organs contaminated with cysts of the tapeworm
Foodborne illnesses (commonly diarrheal diseases)	*Campylobacter* spp., *Escherichia coli*, *Salmonella* spp., *Shigella* spp. and *Trichinella* spp.	animals domesticated for food production (cattle, poultry)	raw and/or undercooked food made from animals
Histoplasmosis	*Histoplasma capsulatum*	birds, bats	inhaling fungi
Influenza	*Influenza A virus*	horses, pigs, domestic and wild birds, wild aquatic mammals such as seals and whales, minks and farmed carnivores	droplets transmitted through air
Leptospirosis	*Leptospira interrogans*	rats, mice, dogs	direct or indirect contact with urine of infected animals
Rabies	*Rabies virus*	commonly - dogs, bats, monkeys, raccoons, foxes, skunks, cattle, wolves, coyotes, mongooses and cats	through saliva by biting, or through scratches from an infected animal
Tularemia	*Francisella tularensis*	lagomorphs (type A) and rodents (type B)	ticks, deer flies, and other insects
Tuberculosis	*Mycobacterium bovis*	infected cattle, deer, llamas, pigs, domestic cats, wild carnivores (foxes, coyotes) and omnivores (possums, mustelids and rodents)	milk, exhaled air, sputum, urine, faeces and pus from infected animals
Trichinosis	*Trichinella spiralis*, *Trichinella britovi*	rodents, pigs, horses, bears, walruses	eating infected meat
Leprosy	*Mycobacterium leprae*	mainly armadillos but mangabey monkeys, rabbits and mice too	any contact with armadillos including urine, faeces and pus from infected animals, the soil infected by armadillo can make you sick too and of course eating infected meat
Chagas disease	*Trypanosoma cruzi*	armadillos, Triatominae (kissing bug)	bite
Toxocariasis	*Toxocara canis* and *Toxocara cati*	dogs, cats	exposure to feces
Toxoplasmosis	*Toxoplasma gondii*	cats, livestock, poultry	exposure to cat feces, and undercooked meat

Partial List of Outbreaks of Associated with Fairs and Petting Zoos

Outbreaks of zoonoses have been traced to human interaction with and exposure to animals at fairs, petting zoos, and other settings. In 2005, the Centers for Disease Control and Prevention (CDC) issued an updated list of recommendations for preventing zoonosis transmission in public settings. The recommendations, developed in conjunction with the National Association of State Public Health Veterinarians, include educational responsibilities of venue operators, limiting public and animal contact, and animal care and management.

In 1988, a person became ill with swine influenza virus (swine flu) and died after visiting the display area of the pig barn at a Wisconsin county fair. Three healthcare personnel treating the case patient also developed flu-like illness with laboratory evidence of swine influenza virus infection. Investigators from the CDC indicated in their final report that the swine flu had been transmitted directly from pig to human host.

In 1994, seven cases of *E. coli* O157:H7 infection were traced to a farm in Leicestershire, United Kingdom. An epidemiological investigation into the outbreak revealed that the strain of *E. coli* O157:H7 isolated from nine animals on the farm was indistinguishable from the strain isolated from human samples. Investigators concluded that the most likely cause of this outbreak was direct human contact with animals.

In 1995, 43 children who had visited a rural farm in Wales became ill with cryptosporidiosis. *Cryptosporidium* was isolated from seven of the ill children. An epidemiological investigation indicated that the source of the children's illness was contact with calves at the farm.

Also in 1995, at least 13 children became ill with *Cryptosporidiosis* after visiting a farm in Dublin, Ireland. In a case-control study, researchers compared the activities of the 13 ill children, or cases, to the activities of 52 out of 55 people who had visited the farm – the controls. The study revealed that illness was significantly associated with playing in the sand in a picnic area beside a stream where animals had access.

In 1997, an *E. coli* O157:H7 outbreak was identified among one child who lived on an open farm and two children who visited the farm during school parties. Two of the three children developed hemolytic-uremic syndrome (HUS). Isolates collected from the three children and from samples taken at the farm were indistinguishable, demonstrating evidence of the link between the farm and the children's illness.

In 1999, what is believed to be the largest outbreak of waterborne *E. coli* O157:H7 illness in United States history occurred at the Washington County, New York fair. The New York State Department of Health identified 781 individuals who were suspected of being infected with either *E. coli* O157:H7 or *Campylobacter jejuni*. An investigation into the outbreak revealed that consumption of beverages purchased from vendors supplied with water drawn from an unchlorinated fairgrounds well was associated with illness. In all, 127 outbreak victims were confirmed ill with *E. coli* O157:H7 infections; 71 were hospitalized, 14 developed HUS, and two died.

In 2000, 51 people became ill with confirmed or suspected *E. coli* O157:H7 infections after visiting a dairy farm in Pennsylvania. Eight children developed HUS. A case-control study among visitors to the dairy was conducted jointly by the CDC, Pennsylvania Department of Health, and the

Montgomery County Health Department. The study's authors concluded that *E. coli* was transmitted to visitors as a result of contamination on animal hides and in the environment.

Also in 2000, 43 visitors to the Medina County fair in Ohio were confirmed ill with *E. coli* O157:H7 infections. An investigation into the outbreak suggested that the water system from which food vendors were supplied was the source of the *E. coli* outbreak. Several months later, five children became ill with *E. coli* infections after attending a "Carnival of Horrors" event held at the Medina County fairgrounds. PFGE analysis of the strains of *E. coli* isolated from members of both outbreaks revealed an indistinguishable pattern, and investigators from the Medina County Health Department and the CDC determined that the Medina County Fairgrounds water distribution system was the source of both *E. coli* outbreaks.

In 2001, an *E. coli* O157:H7 outbreak was traced to exposure in the Cow Palace at the Lorain County Fair in Ohio. CDC investigators identified 23 cases of *E. coli* infection associated with attendance at the Lorain County Fair, with additional secondary cases likely. Two people developed HUS. An investigation revealed *E. coli* contamination on doorways, rails, bleachers, and sawdust. Investigators concluded that the Lorain County Fair was the source of the outbreak.

Wyandot County, Ohio, also reported an *E. coli* O157:H7 outbreak in 2001. Ninety-two *E. coli* infections were reported to the Wyandot County Health Department and the CDC, with 27 cases confirmed using laboratory analysis. Two cases developed HUS. Contact with infected cattle was believed to be the source of the outbreak; however, a specific cause was never identified.

In 2002, 75,000 people became ill with *E. coli* O157:H7 infections after visiting a large agricultural fair in Ontario, Canada. Outbreak investigators conducted a case-control study, which indicated that goats and sheep from a petting zoo were the source of the *E. coli* among fair visitors. Other indications were that the fencing and environment surrounding the petting zoo could have been a source of transmission.

What is believed to be the largest *E. coli* O157:H7 outbreak in Oregon history occurred among attendees at the Lane County fair in 2002. An Oregon Department of Human Services – Health Services investigation led to the belief that the *E. coli* outbreak originated from exposure in the sheep and goat barn. In all, 79 people were confirmed ill with *E. coli* infections as part of the outbreak; 22 were hospitalized, and 12 suffered HUS.

In 2003, fair visitors and animal exhibitors at the Fort Bend County Fair in Texas became ill with *E. coli* O157:H7 infections. An outbreak investigation led to the determination that 25 people had become ill with *E. coli* infections after attending the Fort Bend County Fair; seven people were laboratory-confirmed with *E. coli*, and 5 developed HUS or TTP (thrombotic thrombocytopenic purpura). Investigators isolated a strain of *E. coli* indistinguishable from the outbreak strain from four animal husbandry sites, and found high levels of *E. coli* contamination in both rodeo and animal exhibit areas.

In 2004, a large *E. coli* O157:H7 outbreak occurred among visitors at the 2004 North Carolina State Fair. During its investigation into the outbreak, the North Carolina Department of Health and Human Services (NCDHHS) received over 180 reports of illness, and documented 33 culture-confirmed cases of *E. coli* O157:H7 associated with attendance at the fair, with 15 children developing HUS. In its final investigation report, NCDHHS concluded that the North Carolina

State Fair *E. coli* outbreak had originated at a petting zoo exhibit. The conclusion was supported by a case-control study, environmental sampling, and laboratory analysis of samples collected from the fair and members of the outbreak.

In 2005, a petting zoo that exhibited at two Florida fairs and a festival was traced as the source of an *E. coli* O157:H7 outbreak. Sixty-three people who had visited either the Florida State Fair, the Central Florida Fair, or the Florida Strawberry Festival reported illness to investigators for the Florida Department of Health, including 20 who were culture-confirmed and 7 with HUS. A case-control study revealed that illness was associated with exposure to a petting zoo exhibit present at all three events.

Echinococcosis

Echinococcosis, also called hydatid disease, hydatidosis, or echinococcal disease, is a parasitic disease of tapeworms of the *Echinococcus* type. The two main types of the disease are *cystic echinococcosis* and *alveolar echinococcosis*. Less common forms include *polycystic echinococcosis* and *unicystic echinococcosis*. The disease often starts without symptoms and this may last for years. The symptoms and signs that occur depend on the cyst's location and size. Alveolar disease usually begins in the liver but can spread to other parts of the body, such as the lungs or brain. When the liver is affected the person may have abdominal pain, weight loss, and turn slightly yellow from jaundice. Lung disease may cause pain in the chest, shortness of breath and coughing.

The disease is spread when food or water that contains the eggs of the parasite is eaten or by close contact with an infected animal. The eggs are released in the stool of meat-eating animals that are infected by the parasite. Commonly infected animals include dogs, foxes and wolves. For these animals to become infected they must eat the organs of an animal that contains the cysts such as sheep or rodents. The type of disease that occurs in people depends on the type of *Echinococcus* causing the infection. It is a type of worm infection and is a neglected tropical disease. Diagnosis is usually by ultrasound though computer tomography (CT) or magnetic resonance imaging (MRI) may also be used. Blood tests looking for antibodies against the parasite may be helpful as may biopsy.

Prevention of cystic disease is by treating dogs that may carry the disease and vaccination of sheep. Treatment is often difficult. The cystic disease may be drained through the skin, followed by medication. Sometimes this type of disease is just watched. The alveolar type often needs surgery followed by medications. The medication used is albendazole, which may be needed for years. The alveolar disease may result in death.

The disease occurs in most areas of the world and currently affects about one million people. In some areas of South America, Africa, and Asia up to 10% of the certain populations are affected. In 2010, it caused about 1200 deaths down from 2000 in 1990. The economic cost of the disease is estimated to be around 3 billion USD a year. It is classified as a neglected tropical disease and belongs to the group of diseases known as helminthiasis. It can affect other animals such as pigs, cows and horses.

Classification

The most common form found in humans is cystic echinococcosis (also known as unilocular echinococcosis), which is caused by *Echinococcus granulosus sensu lato*. The second most common form is alveolar echinococcosis (also known as alveolar colloid of the liver, alveolar hydatid disease, alveolococcosis, multilocular echinococcosis, "small fox tapeworm"), which is caused by *Echinococcus multilocularis* and the third is polycystic echinococcosis (also known as human polycystic hydatid disease, neotropical echinococcosis), which is caused by *Echinococcus vogeli* and very rarely, *Echinococcus oligarthrus*. Alveolar and polycystic echinococcosis are rarely diagnosed in humans and are not as widespread as cystic echinococcosis, but polycystic echinococcosis is relatively new on the medical scene and is often left out of conversations dealing with echinococcosis, and alveolar echinococcosis is a serious disease that has not only a significantly high fatality rate, but the potential to become an emerging disease in many countries.

Signs and Symptoms

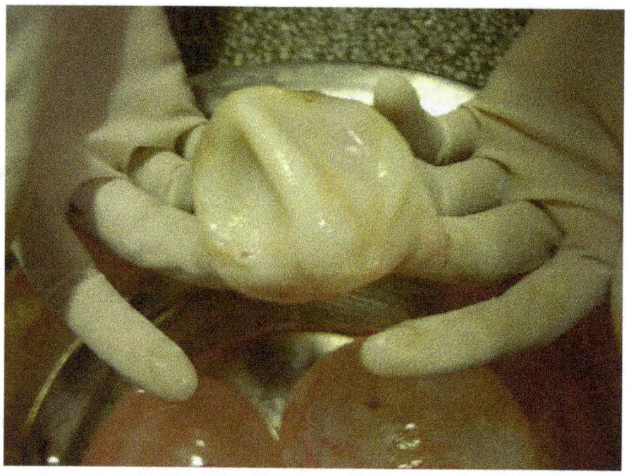

Appearance of a typical cyst at removal

In the human manifestation of the disease, *E. granulosus*, *E. multilocularis*, *E. oligarthrus* and *E. vogeli* are localized in the liver (in 75% of cases), the lungs (in 5–15% of cases) and other organs in the body such as the spleen, brain, heart, and kidneys (in 10–20% of cases). In the patients who are infected with *E. granulosus* and therefore have cystic echinococcosis, the disease develops as a slow-growing mass in the body. These slow-growing masses, often called cysts, are also found in patients that are infected with alveolar and polycystic echinococcosis. The cysts found in those with cystic echinococcosis are usually filled with a clear fluid called hydatid fluid, are spherical, and typically consist of one compartment and are usually only found in one area of the body. While the cysts found in those with alveolar and polycystic echinococcosis are similar to those found in those with cystic echinococcosis, the alveolar and polycystic echinococcosis cysts usually have multiple compartments and have infiltrative as opposed to expansive growth.

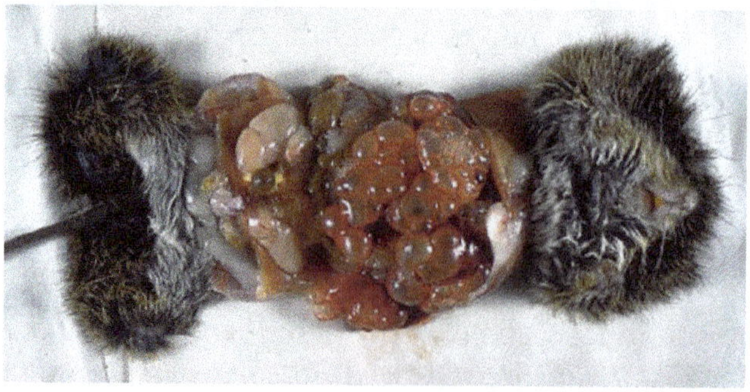

Cysts in a cotton rat

Depending on the location of the cyst in the body, the patient could be asymptomatic even though the cysts have grown to be very large, or be symptomatic even if the cysts are absolutely tiny. If the patient is symptomatic, the symptoms will depend largely on where the cysts are located. For instance, if the patient has cysts in the lungs and is symptomatic, they will have a cough, shortness of breath and/or pain in the chest. On the other hand, if the patient has cysts in the liver and is symptomatic, they will suffer from abdominal pain, abnormal abdominal tenderness, hepatomegaly with an abdominal mass, jaundice, fever and/or anaphylactic reaction. In addition, if the cysts were to rupture while in the body, whether during surgical extraction of the cysts or by trauma to the body, the patient would most likely go into anaphylactic shock and suffer from high fever, pruritus (itching), edema (swelling) of the lips and eyelids, dyspnea, stridor and rhinorrhea.

Unlike intermediate hosts, definitive hosts are usually not hurt very much by the infection. Sometimes, a lack of certain vitamins and minerals can be caused in the host by the very high demand of the parasite.

The incubation period for all species of *Echinococcus* can be months to years, or even decades. It largely depends on the location of the cyst in the body and how fast the cyst is growing.

Cause

Also like many other parasite infections, the course of *Echinococcus* infection is complex. The worm has a life cycle that requires definitive hosts and intermediate hosts. Definitive hosts are normally carnivores such as dogs, while intermediate hosts are usually herbivores such as sheep and cattle. Humans function as accidental hosts, because they are usually a dead end for the parasitic infection cycle.

Hosts

Organism	Definitive Hosts	Intermediate Hosts
E. granulosus	dogs and other canidae	sheep, goats, cattle, camel, buffalo, swine, kangaroos, and other wild herbivores
E. multilocularis	foxes, dogs, other canidae and cats	small rodents
E. vogeli	bush dogs and dogs	rodents
E. oligarthrus	wild felids	small rodents

Life Cycle

An adult worm resides in the small intestine of a definitive host. Gravid proglottids release eggs that are passed in the feces of the definitive host. The egg is then ingested by an intermediate host. The egg then hatches in the small intestine of the intermediate host and releases an oncosphere that penetrates the intestinal wall and moves through the circulatory system into different organs, in particular the liver and lungs. Once it has invaded these organs, the oncosphere develops into a cyst. The cyst then slowly enlarges, creating protoscolices and daughter cysts within the cyst. The definitive host then becomes infected after ingesting the cyst-containing organs of the infected intermediate host. After ingestion, the protoscolices attach to the intestine. They then develop into adult worms and the cycle starts all over again.

Eggs

Echinococcus eggs contain an embryo that is called an oncosphere or hexcanth. The name of this embryo stems from the fact that these embryos have six hooklets. The eggs are passed through the feces of the definitive host and it is the ingestion of these eggs that lead to infection in the intermediate host.

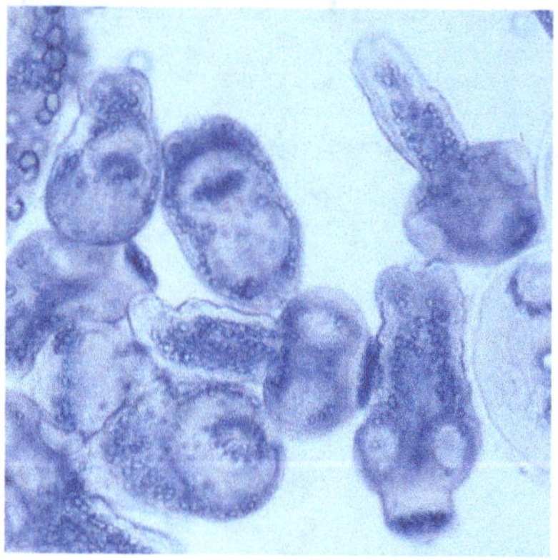

Protoscolices being released from a hydatid cyst

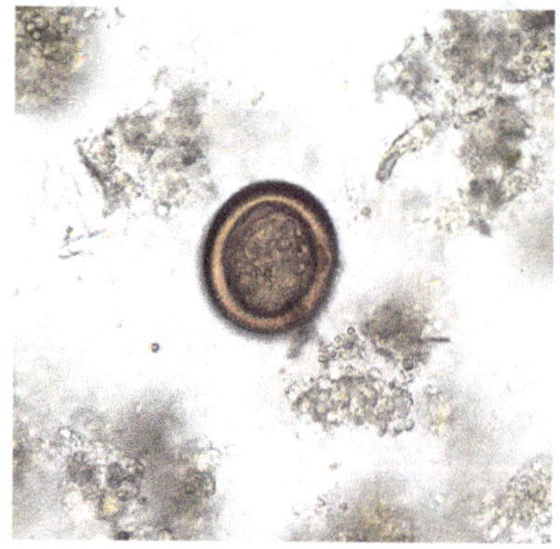

Echinococcus Egg in Feces

Larval/Hydatid Cyst Stage

From the embryo released from an egg develops a *hydatid cyst*, which grows to about 5–10 cm within the first year and is able to survive within organs for years. Cysts sometimes grow to be so large that by the end of several years or even decades, they can contain several liters of fluid. Once a cyst has reached a diameter of 1 cm, its wall differentiates into a thick outer, non-cellular membrane, which covers the thin germinal epithelium. From this epithelium, cells begin to grow within the cyst. These cells then become vacuolated, and are known as brood capsules, which are the parts of the parasite from which protoscolices bud. Often, daughter cysts also form within cysts.

Adult Worm

Echinococcus adult worms develop from protoscolices and are typically 6mm or less in length and have a scolex, neck and typically three proglottids, one of which is immature, another of which is mature and the third of which is gravid (or containing eggs). The scolex of the adult worm contains four suckers and a rostellum that has about 25–50 hooks.

Morphological Differences

The major morphological difference among different species of *Echinococcus* is the length of the tapeworm. *E. granulosus* is approximately 2 to 7 mm while *E. multilocuralis* is often smaller and is 4 mm or less. On the other hand, *E. vogeli* is found to be up to 5.6 mm long and *E. oligarthrus* is found to be up to 2.9 mm long. In addition to the difference in length, there are also differences in the hydatid cysts of the different species. For instance, in *E. multilocularis*, the cysts have an ultra thin limiting membrane and the germinal epithelium may bud externally. Furthermore, *E. granulosus* cysts are unilocular and full of fluid while *E. multilocularis* cysts contain little fluid and are multilocular. For *E. vogeli*, its hydatid cysts are large and are actually polycystic since the germinal membrane of the hydatid cyst actually proliferates both inward, to create septa that divide the hydatid into sections, and outward, to create new cysts. Like *E. granulosus* cysts, *E. vogeli* cysts are filled with fluid.

Transmission

As one can see from the life cycles illustrated above, all disease-causing species of *Echinococcus* are transmitted to intermediate hosts via the ingestion of eggs and are transmitted to definitive hosts by means of eating infected, cyst-containing organs. Humans are accidental intermediate hosts that become infected by handling soil, dirt or animal hair that contains eggs.

While there are no biological or mechanical vectors for the adult or larval form of any *Echinococcus* species, coprophagic flies, carrion birds and arthropods can act as mechanical vectors for the eggs.

Diagnosis

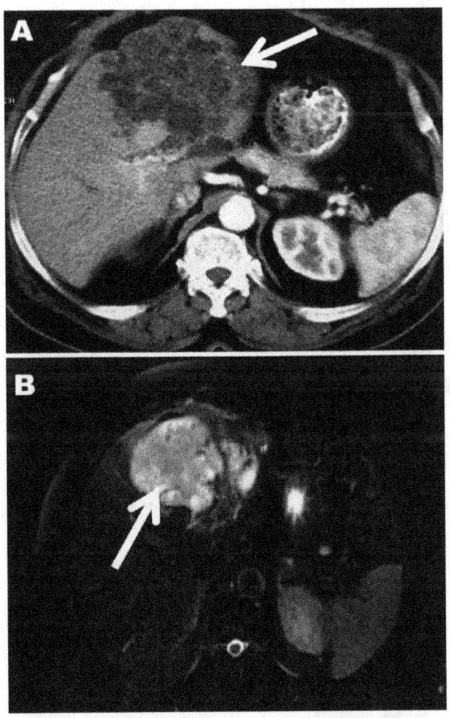

Computed tomography (A) and magnetic resonance (B) images of the liver of a 72-year-old man from French Guiana with polycystic echinococcosis affecting the left side of the liver. White arrows indicate the multicystic liver lesion.

Cystic

A formal diagnose of any type of echinococcosis requires a combination of tools that involve imaging techniques, histopathology, or nucleic acid detection and serology. For cystic echinococcosis diagnosis, imaging is the main method—while serology tests (such as indirect hemogglutination, ELISA (enzyme linked immunosorbent assay), immunoblots or latex agglutination) that use antigens specific for *E. granulosus* verify the imaging results. The imaging technique of choice for cystic echinococcosis is ultrasonography, since it is not only able to visualize the cysts in the body's organs, but it is also inexpensive, non-invasive and gives instant results. In addition to ultrasonography, both MRI and CT scans can and are often used although an MRI is often preferred to CT scans when diagnosing cystic echinococcosis since it gives better visualization of liquid areas within the tissue.

Alveolar

As with cystic echinococcosis, ultrasonography is the imaging technique of choice for alveolar echinococcosis and is usually complemented by CT scans since CT scans are able to detect the largest number of lesions and calcifications that are characteristic of alveolar echinococcosis. MRIs are also used in combination with ultrasonography though CT scans are preferred. Like cystic echinococcosis, imaging is the major method used for the diagnosis of alveolar echinococcosis while the same types of serologic tests (except now specific for *E. multilocularis* antigens) are used to verify the imaging results. It is also important to note that serologic tests are more valuable for the diagnosis of alveolar echinococcosis than for cystic echinococcosis since they tend to be more reliable for alveolar echinococcosis since more antigens specific for *E. multilocularis* are available. In addition to imaging and serology, identification of *E. multilocularis* infection via PCR or a histological examination of a tissue biopsy from the patient is another way to diagnose alveolar echinococcosis.

Polycystic

Similar to the diagnosis of alveolar echinococcosis and cystic echinococcosis, the diagnosis of polycystic echinococcosis uses imaging techniques, in particular ultrasonography and CT scans, to detect polycystic structures within the patient's body. However, imaging is not the preferred method of diagnosis since the method that is currently considered the standard is the isolation of protoscoleces during surgery or after the patient's death and the identification of definitive features of *E. oligarthrus* and *E. vogeli* in these isolated protoscoleces. This is the main way that PE is diagnosed, but some current studies show that PCR may identify *E. oligarthrus* and *E. vogeli* in patients' tissues. The only drawback of using PCR to diagnose polycystic echinococcosis is that there aren't many genetic sequences that can be used for PCR that are specific only *E. oligarthrus* or *E. vogeli*.

Prevention

Boiling sheep livers containing hydatid cysts for 30 minutes kills the cestode larvae – a simple method for prevention in remote areas

Cystic Echinococcosis

There are several different strategies that are currently being used to prevent and control cystic echinococcosis (CE). Most of these various methods try to prevent and control CE by targeting the major risk factors for the disease and the way it is transmitted. For instance, health education programs focused on cystic echinococcosis and its agents, and improved water sanitation attempt to target poor education and poor drinking water sources, which are both risk factors for contracting echinococcosis. Furthermore, since humans often come into contact with *Echinococcus* eggs via touching contaminated soil, animal feces and animal hair, another prevention strategy is improved hygiene. In addition to targeting risk factors and transmission, control and prevention strategies of cystic echinococcosis also aim at intervening at certain points of the parasite's life cycle, in particular, the infection of hosts (i.e. especially dogs) that reside with or near humans. For example, many countries endemic to echinococcosis have implemented programs geared at de-worming dogs and vaccinating dogs and other livestock, such as sheep, that also act as hosts for *E. granulosus*. Proper disposal of carcasses and offal after home slaughter is difficult in poor and remote communities and therefore dogs readily have access to offal from livestock, thus completing the parasite cycle of *Echinococcus granulosus* and putting communities at risk of cystic echinococcosis. Boiling livers and lungs that contain hydatid cysts during 30 minutes has been proposed as a simple, efficient and energy- and time-saving way to kill the infectious larvae.

Alveolar Echinococcosis

A number of strategies are geared towards prevention and control of alveolar echinococcosis—most of which are similar to those for cystic echinococcosis. For instance, health education programs, improved water sanitation, improved hygiene and de-worming of hosts (particularly red foxes) are all effective to prevent and control the spread of alveolar echinococcosis. Unlike cystic echinococcosis, however, where there is a vaccine against *E. granulosus*, there is currently no canidae or livestock vaccine against *E. multilocularis*.

Polycystic Echinococcosis

While a number of control and prevention strategies deal with cystic and alveolar echinococcosis, there are few methods to control and prevent polycystic echinococcosis. This is probably due to the fact that polycystic echinococcosis is restricted to Central and South America, and that the way that humans become accidental hosts of *E. oligarthrus* and *E. vogeli* is still not completely understood.

Human Vaccines

Currently there are no human vaccines against any form of echinococcosis. However, there are studies being conducted that are looking at possible vaccine candidates for an effective human vaccine against echinococcosis.

Treatment

Cystic

For simple cases of cystic echinococcosis, the most common form of treatment is open surgical removal of the cysts combined with chemotherapy using albendazole and/or mebendazole before and after surgery. However, if there are cysts in multiple organs or tissues, or the cysts are in risky locations, surgery becomes impractical. For inoperable cases such as these, chemotherapy and/or PAIR (puncture-aspiration-injection-reaspiration) become alternative options of treatment. In the case of alternative treatment using just chemotherapy, albendazole is preferred twice a day for 1–5 months. An alternative to albendazole is mebendazole for at least 3 to 6 months. The other alternative to surgery is PAIR with chemotherapy. PAIR is a minimally invasive procedure that involves three steps: puncture and needle aspiration of the cyst, injection of a scolicidal solution for 20–30 min, and cyst-re-aspiration and final irrigation. Patients who undergo PAIR typically take albendazole or mebendazole from 7 days before the procedure until 28 days after the procedure. While open surgery still remains as the standard for cystic echinococcosis treatment, there have been a number of studies that suggest that PAIR with chemotherapy is more effective than surgery in terms of disease recurrence, and morbidity and mortality. In addition to the three above mentioned treatments, there is currently research and studies looking at new treatment involving percutaneous thermal ablation (PTA) of the germinal layer in the cyst by means of a radiofrequency ablation device. This form of treatment is still relatively new and requires much more testing before being widely used. An alternative to open surgery is laparoscopic surgery, which provides excellent cure rates with minimal morbidity and mortality.

Alveolar

For alveolar echinococcosis, surgical removal of cysts combined with chemotherapy (using albendazole and/or mebendazole) for up to two years after surgery is the only sure way to completely cure the disease. However, in inoperable cases, chemotherapy by itself can also be used. In treatment using just chemotherapy, one could use either mebendazole in three doses or albendazole in two doses. Since chemotherapy on its own is not guaranteed to completely rid the patient of disease, patients are often kept on the drugs for extended periods of times (i.e. more than 6 months, years). In addition to surgery and chemotherapy, liver transplants are being looked into as a form of treatment for alveolar echinococcosis although it is seen as incredibly risky since it often leads to echinococcosis re-infection in the patient afterwards.

Polycystic

Since polycystic echinococcosis is constrained to such a particular area of the world and is not well described or found in many people, treatment of polycystic echinococcosis is less defined than that of cystic and alveolar echinococcosis. While surgical removal of cysts was the treatment of choice for the previous two types of echinococcosis, chemotherapy is the recommended treatment approach for polycystic echinococcosis. While albendazole is the preferred drug, mebendazole can also be used if the treatment is to be for an extended period of time. Only if chemotherapy fails or if the lesions are very small is surgery advised.

Epidemiology

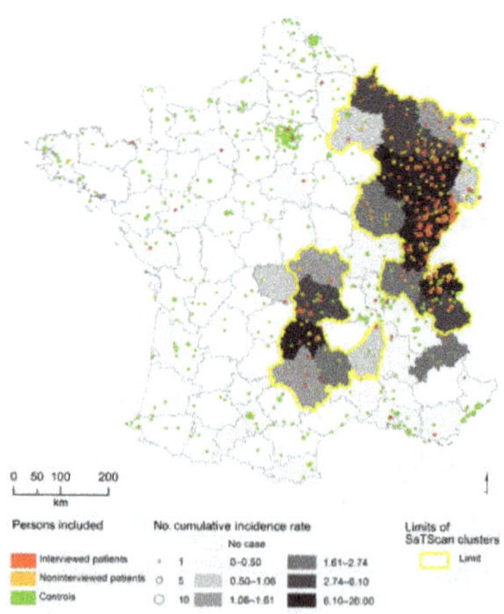

Location of patients, controls, and areas in France where persons are at risk for alveolar echinococcosis. See legend below the image.

Regions

Very few countries are considered to be completely free of *E. granulosus*. Areas of the world where there is a high rate of infection often coincide with rural, grazing areas where dogs are able to ingest organs from infected animals.

E. multilocularis mainly occurs in the Northern hemisphere, including central Europe and the northern parts of Europe, Asia, and North America. However, its distribution was not always like this. For instance, until the end of the 1980s, *E. multilocularis* endemic areas in Europe were known to exist only in France, Switzerland, Germany, and Austria. But during the 1990s and early 2000s, there was a shift in the distribution of *E. multilocularis* as the infection rate of foxes escalated in certain parts of France and Germany. As a result, several new endemic areas were found in Switzerland, Germany, and Austria and surrounding countries such as the Netherlands, Belgium, Luxembourg, Poland, the Czech Republic, the Slovak Republic, and Italy. While alveolar echinococcosis is not extremely common, it is believed that in the coming years, it will be an emerging or re-emerging disease in certain countries as a result of *E. multilocularis*' ability to spread.

Unlike the previous two species of *Echinococcus*, *E. vogeli* and *E. oligarthrus* are limited to Central and South America. Furthermore, infections by *E. vogeli* and *E. oligarthrus* (polycystic echinococcosis) are considered to be the rarest form of echinococcosis.

Deaths

As of 2010 it caused about 1200 deaths, down from 2000 in 1990.

Largest Discovery

The Largest hydatid cyst contained 37 litres of fluid and was removed from 80 year old Abdullah Bakhet Al Hagawe by Dr. Abdul Rahman Arishi and Dr. Mohammad Shahid Hussain Khan, at the King Fahd Central Hospital in Jizan, Saudi Arabia, on 12 January 2005. The main cysts had multiple daughter cysts attached and was found in his abdomen.

History

Echinococcosis is a disease that has been recognized by humans for centuries. There has been mention of it in the Talmud. It was also recognized by ancient scholars such as Hippocrates, Aretaeus, Galen and Rhazes. Although echinococcosis has been well known for the past two thousand years, it wasn't until the past couple of hundred years that real progress was made in determining and describing its parasitic origin. The first step towards figuring out the cause of echinococcosis occurred during the 17th century when Francesco Redi illustrated that the hydatid cysts of echinococcosis were of "animal" origin. Then, in 1766, Pierre Simon Pallas predicted that these hydatid cysts found in infected humans were actually larval stages of tapeworms. A few decades afterwards, in 1782, Goeze accurately described the cysts and the tapeworm heads while in 1786, *E. granulosus* was accurately described by Batsch. Half a century later, during the 1850s, Karl von Siebold showed through a series of experiments that *Echinococcus* cysts do cause adult tapeworms in dogs. Shortly after this, in 1863, *E. multilocularis* was identified by Rudolf Leuckart. Then, during the early to mid 1900s, the more distinct features of *E. granulosus and E. multilocularis*, their life cycles and how they cause disease were more fully described as more and more people began researching and performing experiments and studies. While *E. granulosus* and *E. multilocularis* were both linked to human echinococcosis before or shortly after the 20th century, it wasn't until the mid 1900s that *E. oligarthrus* and *E. vogeli* were identified as and shown as being causes of human echinococcosis.

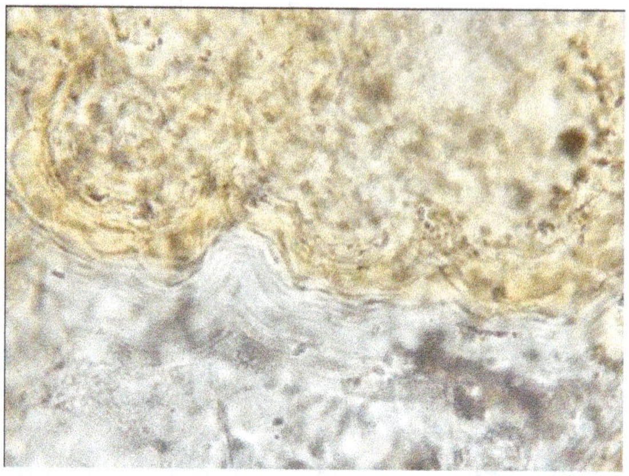

Laminated layer of a cyst found in a corpse in a grave from the late Roman period in France, interpreted as a probable hydatidosis cyst

Two calcified objects recovered from a 3rd to 4th-century grave of an adolescent in Amiens (Northern France) were interprated as probable hydatid cysts. A study of remains from two 8,000-year-old cemeteries in Siberia showed presence of echinococcosis.

References

- Smith, David F.; Diack, H. Lesley and Pennington, T. Hugh (2005) Food Poisoning, Policy and Politics : Corned Beef and Typhoid in Britain in the 1960s. Boydell Press. ISBN 1-84383-138-4
- John, David T.; William Petri, William A.; Markell, Edward K.; Voge, Marietta (January 2006). "7: The Cestodes: Echinococcus granulosus, E. multiloularis and E. vogeli (Hyatid Disease)". Markell and Voge's Medical Parasitology (9th ed.). Elsevier Health Sciences. pp. 224–231. ISBN 0-7216-4793-6.
- Mandell, Gerald L.; Bennett, John Eugene; Dolin, Raphael, eds. (2010). "Ch. 290". Mandell, Douglas, and Bennett's Principles and Practice of Infectious Diseases (7th ed.). Elsevier. ISBN 0443068399.
- "Meat Produced by Advanced Meat/Bone Separation Machinery and Meat Recovery (AMR) Systems". Federal Register. U.S. National Archive and Records Administration. Retrieved 3 July 2016.
- Kirk, Martyn; et al. "Foodborne illness in Australia: Annual incidence circa 2010 pp7-9". Australia Department of Health. Australian National University. Retrieved 13 September 2015.
- Messenger AM, Barnes AN, Gray GC (2014). "Reverse zoonotic disease transmission (zooanthroponosis): a systematic review of seldom-documented human biological threats to animals.". PLOS ONE. 9 (2): e89055. doi:10.1371/journal.pone.0089055. PMID 24586500. Retrieved 18 December 2014.
- Information in this table is largely compiled from: World Health Organization. "Zoonoses and the Human-Animal-Ecosystems Interface". Retrieved 21 December 2014.
- "Foodborne Illness - Frequently Asked Questions". US Centers for Disease Control and Prevention. Archived from the original on March 3, 2011. Retrieved 3 July 2016.

5

Food Preservation: Tools and Techniques

Food preservation techniques stop food from spoilage and increase its shelf-life. Traditional methods of preservation like pickling of vegetables and salting or curing of meat are some of the earliest attempts at storing food for difficult times. The chapter discusses methods like pasteurization, addition of preservatives, electroporation, pascalization, food irradiation, biopreservation and food storage. The chapter strategically encompasses and incorporates the major components and key concepts of the food industry, providing a complete understanding.

Food Preservation

Food preservation involves preventing the growth of bacteria, fungi (such as yeasts), or other micro-organisms (although some methods work by introducing benign bacteria or fungi to the food), as well as slowing the oxidation of fats that cause rancidity. Food preservation may also include processes that inhibit visual deterioration, such as the enzymatic browning reaction in apples after they are cut during food preparation.

Many processes designed to preserve food will involve a number of food preservation methods. Preserving fruit by turning it into jam, for example, involves boiling (to reduce the fruit's moisture content and to kill bacteria, etc.), sugaring (to prevent their re-growth) and sealing within an airtight jar (to prevent recontamination). Some traditional methods of preserving food have been shown to have a lower energy input and carbon footprint, when compared to modern methods. However, some methods of food preservation are known to create carcinogens, and in 2015, the International Agency for Research on Cancer of the World Health Organization classified processed meat, i.e. meat that has undergone salting, curing, fermenting, and smoking, as "carcinogenic to humans".

Maintaining or creating nutritional value, texture and flavor is an important aspect of food preservation, although, historically, some methods drastically altered the character of the food being preserved. In many cases these changes have come to be seen as desirable qualities – cheese, yogurt and pickled onions being common examples.

Traditional Techniques

New techniques of food preservation became available to the home chef from the dawn of agriculture until the Industrial Revolution.

Drying

Drying is one of the oldest techniques used to hamper the decomposition of food products. As early as 12,000 B.C., Middle Eastern and Oriental cultures were drying foods using the power of the sun.

Vegetables and fruit are naturally dried by the sun and wind, but "still houses" were built in areas that did not have enough sunlight to dry things. A fire would be built inside the building to provide the heat to dry the various fruits, vegetables, and herbs.

Cooling

Cooling preserves foods by slowing down the growth and reproduction of micro-organisms and the action of enzymes that cause food to rot. The introduction of commercial and domestic refrigerators drastically improved the diets of many in the Western world by allowing foods such as fresh fruit, salads and dairy products to be stored safely for longer periods, particularly during warm weather.

Freezing

Freezing is also one of the most commonly used processes, both commercially and domestically, for preserving a very wide range of foods, including prepared foods that would not have required freezing in their unprepared state. For example, potato waffles are stored in the freezer, but potatoes themselves require only a cool dark place to ensure many months' storage. Cold stores provide large-volume, long-term storage for strategic food stocks held in case of national emergency in many countries.

Boiling

Boiling liquid food items can kill any existing microbes. Milk and water are often boiled to kill any harmful microbes that may be present in them.

Heating

Heating to temperatures which are sufficient to kill microorganisms inside the food is a method used with perpetual stews. Milk is also boiled before storing to kill many microorganisms.

Salting

Bag of Prague powder #1, also known as "curing salt" or "pink salt". It is typically a combination of salt and sodium nitrite, with the pink color added to distinguish it from ordinary salt.

Salting or curing draws moisture from the meat through a process of osmosis. Meat is cured with salt or sugar, or a combination of the two. Nitrates and nitrites are also often used to cure meat and contribute the characteristic pink color, as well as inhibition of *Clostridium botulinum*. It was a main method of preservation in medieval times and around the 1700s.

Sugaring

The earliest cultures have used sugar as a preservative, and it was commonplace to store fruit in honey. Similar to pickled foods, sugar cane was brought to Europe through the trade routes. In northern climates without sufficient sun to dry foods, preserves are made by heating the fruit with sugar. "Sugar tends to draw water from the microbes (plasmolysis). This process leaves the microbial cells dehydrated, thus killing them. In this way, the food will remain safe from microbial spoilage." Sugar is used to preserve fruits, either in an anti-microbial syrup with fruit such as apples, pears, peaches, apricots and plums, or in crystallized form where the preserved material is cooked in sugar to the point of crystallization and the resultant product is then stored dry. This method is used for the skins of citrus fruit (candied peel), angelica and ginger. Also sugaring can be used in jam jellies.

Smoking

Smoking is used to lengthen the shelf life of perishable food items. This effect is achieved by exposing the food to smoke from burning plant materials such as wood. Smoke deposits a number of pyrolysis products onto the food, including the phenols syringol, guaiacol and catechol. These compounds aid in the drying and preservation of meats and other foods. Most commonly subjected to this method of food preservation are meats and fish that have undergone curing. Fruits and vegetables like paprika, cheeses, spices, and ingredients for making drinks such as malt and tea leaves are also smoked, but mainly for cooking or flavoring them. It is one of the oldest food preservation methods, which probably arose after the development of cooking with fire.

Smoked ham in a smokehouse in Schleswig-Holstein, Germany

Pickling

Pickling is a method of preserving food in an edible anti-microbial liquid. Pickling can be broadly classified into two categories: chemical pickling and fermentation pickling.

In chemical pickling, the food is placed in an edible liquid that inhibits or kills bacteria and other micro-organisms. Typical pickling agents include brine (high in salt), vinegar, alcohol, and vegetable oil, especially olive oil but also many other oils. Many chemical pickling processes also involve heating or boiling so that the food being preserved becomes saturated with the pickling agent. Common chemically pickled foods include cucumbers, peppers, corned beef, herring, and eggs, as well as mixed vegetables such as piccalilli.

In fermentation pickling, the food itself produces the preservation agent, typically by a process that produces lactic acid. Fermented pickles include sauerkraut, nukazuke, kimchi, surströmming, and curtido. Some pickled cucumbers are also fermented.

Lye

Sodium hydroxide (lye) makes food too alkaline for bacterial growth. Lye will saponify fats in the food, which will change its flavor and texture. Lutefisk uses lye in its preparation, as do some olive recipes. Modern recipes for century eggs also call for lye.

Canning

Canning involves cooking food, sealing it in sterile cans or jars, and boiling the containers to kill or weaken any remaining bacteria as a form of sterilization. It was invented by the French confectioner Nicolas Appert. By 1806, this process was used by the French Navy to preserve meat, fruit, vegetables, and even milk. Although Appert had discovered a new way of preservation, it wasn't understood until 1864 when Louis Pasteur found the relationship between microorganisms, food spoilage, and illness.

Preserved food

Foods have varying degrees of natural protection against spoilage and may require that the final step occur in a pressure cooker. High-acid fruits like strawberries require no preservatives to can and only a short boiling cycle, whereas marginal vegetables such as carrots require longer boiling and addition of other acidic elements. Low-acid foods, such as vegetables and meats, require pressure canning. Food preserved by canning or bottling is at immediate risk of spoilage once the can or bottle has been opened.

Lack of quality control in the canning process may allow ingress of water or micro-organisms. Most such failures are rapidly detected as decomposition within the can causes gas production and the can will swell or burst. However, there have been examples of poor manufacture (underprocessing) and poor hygiene allowing contamination of canned food by the obligate anaerobe *Clostridium botulinum*, which produces an acute toxin within the food, leading to severe illness or death. This organism produces no gas or obvious taste and remains undetected by taste or smell. Its toxin is denatured by cooking, however. Cooked mushrooms, handled poorly and then canned, can support the growth of Staphylococcus aureus, which produces a toxin that is not destroyed by canning or subsequent reheating.

Jellying

Food may be preserved by cooking in a material that solidifies to form a gel. Such materials include gelatin, agar, maize flour, and arrowroot flour. Some foods naturally form a protein gel when cooked, such as eels and elvers, and sipunculid worms, which are a delicacy in Xiamen, in the Fujian province of the People's Republic of China. Jellied eels are a delicacy in the East End of London, where they are eaten with mashed potatoes. Potted meats in aspic (a gel made from gelatine and clarified meat broth) were a common way of serving meat off-cuts in the UK until the 1950s. Many jugged meats are also jellied.

A traditional British way of preserving meat (particularly shrimp) is by setting it in a pot and sealing it with a layer of fat. Also common is potted chicken liver; compare pâté.

Jugging

Meat can be preserved by jugging. Jugging is the process of stewing the meat (commonly game or fish) in a covered earthenware jug or casserole. The animal to be jugged is usually cut into pieces, placed into a tightly-sealed jug with brine or gravy, and stewed. Red wine and/or the animal's own blood is sometimes added to the cooking liquid. Jugging was a popular method of preserving meat up until the middle of the 20th century.

Burial

Burial of food can preserve it due to a variety of factors: lack of light, lack of oxygen, cool temperatures, pH level, or desiccants in the soil. Burial may be combined with other methods such as salting or fermentation. Most foods can be preserved in soil that is very dry and salty (thus a desiccant) such as sand, or soil that is frozen.

Many root vegetables are very resistant to spoilage and require no other preservation than storage in cool dark conditions, for example by burial in the ground, such as in a storage clamp. Century eggs are created by placing eggs in alkaline mud (or other alkaline substance), resulting in their "inorganic" fermentation through raised pH instead of spoiling. The fermentation preserves them and breaks down some of the complex, less flavorful proteins and fats into simpler, more flavorful ones. Cabbage was traditionally buried in the fall in northern farms in the U.S. for preservation. Some methods keep it crispy while other methods produce sauerkraut. A similar process is used in the traditional production of kimchi. Sometimes meat is buried under conditions that cause preservation. If buried on hot coals or ashes, the heat can kill pathogens, the dry ash can desiccate, and the earth can block oxygen and further contamination. If buried where the earth is very cold, the earth acts like a refrigerator.

In Orissa, India, it is practical to store rice by burying it underground. This method helps to store for three to six months during the dry season.

Curing

The earliest form of curing was dehydration. To accelerate this process, salt is usually added. In the culinary world, it was common to choose raw salts from various sources (rock salt, sea salt, etc.). More modern "examples of salts that are used as preservatives include sodium chloride (NaCl), sodium nitrate ($NaNO_3$) and sodium nitrite ($NaNO_2$). Even at mild concentrations (up to 2%), sodium chloride, found in many food products, is capable of neutralizing the antimicrobial character of natural compounds."

Fermentation

Some foods, such as many cheeses, wines, and beers, use specific micro-organisms that combat spoilage from other less-benign organisms. These micro-organisms keep pathogens in check by creating an environment toxic for themselves and other micro-organisms by producing acid or alcohol. Methods of fermentation include, but are not limited to, starter micro-organisms, salt, hops, controlled (usually cool) temperatures and controlled (usually low) levels of oxygen. These methods are used to create the specific controlled conditions that will support the desirable organisms that produce food fit for human consumption.

Fermentation is the microbial conversion of starch and sugars into alcohol. Not only can fermentation produce alcohol, but it can also be a valuable preservation technique. Fermentation can also make foods more nutritious and palatable. For example, drinking water in the Middle Ages was dangerous because it often contained pathogens that could spread disease. When the water is made into beer, the boiling during the brewing process kills any bacteria in the water that could make people sick. Additionally, the water now has the nutrients from the barley and other ingredients, and the microorganisms can also produce vitamins as they ferment.

Industrial/Modern Techniques

Techniques of food preservation were developed in research laboratories for commercial applications.

Pasteurization

Pasteurization is a process for preservation of liquid food. It was originally applied to combat the souring of young local wines. Today, the process is mainly applied to dairy products. In this method, milk is heated at about 70 °C for 15 to 30 seconds to kill the bacteria present in it and cooling it quickly to 10 °C to prevent the remaining bacteria from growing. The milk is then stored in sterilized bottles or pouches in cold places. This method was invented by Louis Pasteur, a French chemist, in 1862.

Vacuum Packing

Vacuum-packing stores food in a vacuum environment, usually in an air-tight bag or bottle. The vacuum environment strips bacteria of oxygen needed for survival. Vacuum-packing is commonly used for storing nuts to reduce loss of flavor from oxidization. A major drawback to vacuum packaging, at the consumer level, is that vacuum sealing can deform contents and rob certain foods, such as cheese, of its flavor.

Artificial food Additives

Preservative food additives can be *antimicrobial*, which inhibit the growth of bacteria or fungi, including mold, or *antioxidant*, such as oxygen absorbers, which inhibit the oxidation of food constituents. Common antimicrobial preservatives include calcium propionate, sodium nitrate, sodium nitrite, sulfites (sulfur dioxide, sodium bisulfite, potassium hydrogen sulfite, etc.) and disodium EDTA. Antioxidants include BHA and BHT. Other preservatives include formaldehyde (usually in solution), glutaraldehyde (kills insects), ethanol, and methylchloroisothiazolinone.

Irradiation

Irradiation of food is the exposure of food to ionizing radiation. The two types of ionizing radiation used are beta particles (high-energy electrons) and gamma rays (emitted from radioactive sources as cobalt-60 or cesium-137). Treatment effects include killing bacteria, molds, and insect pests, reducing the ripening and spoiling of fruits, and at higher doses inducing sterility. The technology may be compared to pasteurization; it is sometimes called “cold pasteurization”, as the product is not heated.

The irradiation process is not directly related to nuclear energy, but does use radioactive isotopes produced in nuclear reactors. Cobalt-60, for example does not occur naturally and can only be produced through neutron bombardment of cobalt-59. Ionizing radiation at high energy levels is hazardous to life (hence its usefulness in sterilisation); for this reason, irradiation facilities have a heavily shielded irradiation room where the process takes place. Radiation safety procedures are used to ensure that neither the workers in such facilities nor the environment receives any radiation dose above administrative limits. Irradiated food does not and cannot become radioactive, and national and international expert bodies have declared food irradiation as wholesome. However, the wholesomeness of consuming such food is disputed by opponents and consumer organizations. National and international expert bodies have declared food irradiation as “wholesome”; organizations of the United Nations, such as the World Health Organization and Food and Agriculture Organization, endorse food irradiation. International legislation on whether food may be irradiated or not varies worldwide from no regulation to full banning. Irradiation may allow lower-quality or contaminated foods to be rendered marketable.

Approximately 500,000 tons of food items are irradiated per year worldwide in over 40 countries. These are mainly spices and condiments with an increasing segment of fresh fruit irradiated for fruit fly quarantine.

Pulsed Electric Field Electroporation

Pulsed electric field (PEF) electroporation is a method for processing cells by means of brief pulses of a strong electric field. PEF holds potential as a type of low-temperature alternative pasteurization process for sterilizing food products. In PEF processing, a substance is placed between two electrodes, then the pulsed electric field is applied. The electric field enlarges the pores of the cell membranes, which kills the cells and releases their contents. PEF for food processing is a developing technology still being researched. There have been limited industrial applications of PEF processing for the pasteurization of fruit juices. To date, several PEF treated juices are available on the market in Europe. Furthermore, for several years a juice pasteurization application in the

US has used PEF. For cell disintegration purposes especially potato processors show great interest in PEF technology as an efficient alternative for their preheaters. Potato applications are already operational in the US and Canada. There are also commercial PEF potato applications in various countries in Europe, as well as in Australia, India and China.

Modified Atmosphere

Modifying atmosphere is a way to preserve food by operating on the atmosphere around it. Salad crops that are notoriously difficult to preserve are now being packaged in sealed bags with an atmosphere modified to reduce the oxygen (O_2) concentration and increase the carbon dioxide (CO_2) concentration. There is concern that, although salad vegetables retain their appearance and texture in such conditions, this method of preservation may not retain nutrients, especially vitamins. There are two methods for preserving grains with carbon dioxide. One method is placing a block of dry ice in the bottom and filling the can with the grain. Another method is purging the container from the bottom by gaseous carbon dioxide from a cylinder or bulk supply vessel.

Carbon dioxide prevents insects and, depending on concentration, mold and oxidation from damaging the grain. Grain stored in this way can remain edible for approximately five years.

Nitrogen gas (N_2) at concentrations of 98% or higher is also used effectively to kill insects in the grain through hypoxia. However, carbon dioxide has an advantage in this respect, as it kills organisms through hypercarbia and hypoxia (depending on concentration), but it requires concentrations of above 35%, or so. This makes carbon dioxide preferable for fumigation in situations where a hermetic seal cannot be maintained.

Controlled Atmospheric Storage (CA): "CA storage is a non-chemical process. Oxygen levels in the sealed rooms are reduced, usually by the infusion of nitrogen gas, from the approximate 21 percent in the air we breathe to 1 percent or 2 percent. Temperatures are kept at a constant 0 to 2 °C (32 to 36 °F). Humidity is maintained at 95 percent and carbon dioxide levels are also controlled. Exact conditions in the rooms are set according to the apple variety. Researchers develop specific regimens for each variety to achieve the best quality. Computers help keep conditions constant." "Eastern Washington, where most of Washington's apples are grown, has enough warehouse storage for 181 million boxes of fruit, according to a report done in 1997 by managers for the Washington State Department of Agriculture Plant Services Division. The storage capacity study shows that 67 percent of that space —enough for 121,008,000 boxes of apples — is CA storage."

Air-tight storage of grains (sometimes called hermetic storage) relies on the respiration of grain, insects, and fungi that can modify the enclosed atmosphere sufficiently to control insect pests. This is a method of great antiquity, as well as having modern equivalents. The success of the method relies on having the correct mix of sealing, grain moisture, and temperature.

A patented process uses fuel cells to exhaust and automatically maintain the exhaustion of oxygen in a shipping container, containing, for example, fresh fish.

Nonthermal Plasma

This process subjects the surface of food to a "flame" of ionized gas molecules, such as helium or nitrogen. This causes micro-organisms to die off on the surface.

High-pressure Food Preservation

High-pressure food preservation or pascalization refers to the use of a food preservation technique that makes use of high pressure. "Pressed inside a vessel exerting 70,000 pounds per square inch (480 MPa) or more, food can be processed so that it retains its fresh appearance, flavor, texture and nutrients while disabling harmful microorganisms and slowing spoilage." By 2005, the process was being used for products ranging from orange juice to guacamole to deli meats and widely sold.

Biopreservation

Biopreservation is the use of natural or controlled microbiota or antimicrobials as a way of preserving food and extending its shelf life. Beneficial bacteria or the fermentation products produced by these bacteria are used in biopreservation to control spoilage and render pathogens inactive in food. It is a benign ecological approach which is gaining increasing attention.

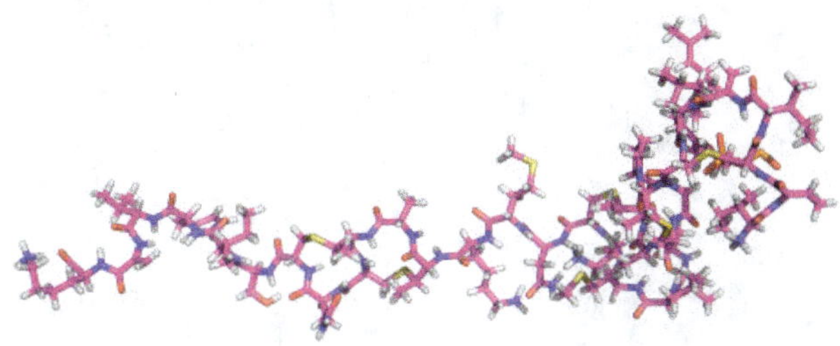

3D stick model of nisin. Some lactic acid bacteria manufacture nisin. It is a particularly effective preservative.

Of special interest are lactic acid bacteria (LAB). Lactic acid bacteria have antagonistic properties that make them particularly useful as biopreservatives. When LABs compete for nutrients, their metabolites often include active antimicrobials such as lactic acid, acetic acid, hydrogen peroxide, and peptide bacteriocins. Some LABs produce the antimicrobial nisin, which is a particularly effective preservative.

These days, LAB bacteriocins are used as an integral part of hurdle technology. Using them in combination with other preservative techniques can effectively control spoilage bacteria and other pathogens, and can inhibit the activities of a wide spectrum of organisms, including inherently resistant Gram-negative bacteria.

Hurdle Technology

Hurdle technology is a method of ensuring that pathogens in food products can be eliminated or controlled by combining more than one approach. These approaches can be thought of as "hurdles" the pathogen has to overcome if it is to remain active in the food. The right combination of hurdles can ensure all pathogens are eliminated or rendered harmless in the final product.

Hurdle technology has been defined by Leistner (2000) as an intelligent combination of hurdles that secures the microbial safety and stability as well as the organoleptic and nutritional quality and the economic viability of food products. The organoleptic quality of the food refers to its sensory properties, that is its look, taste, smell, and texture.

Examples of hurdles in a food system are high temperature during processing, low temperature during storage, increasing the acidity, lowering the water activity or redox potential, and the presence of preservatives or biopreservatives. According to the type of pathogens and how risky they are, the intensity of the hurdles can be adjusted individually to meet consumer preferences in an economical way, without sacrificing the safety of the product.

Tools and Techniques of Food Preservation

Pasteurization

Pasteurization or pasteurisation is a process that kills microbes (mainly bacteria) in food and drink, such as milk, juice, canned food, and others.

Cream pasteurizing and cooling coils at Murgon Butter Factory, 1939

It was invented by French scientist Louis Pasteur during the nineteenth century. In 1864 Pasteur discovered that heating beer and wine was enough to kill most of the bacteria that caused spoilage, preventing these beverages from turning sour. The process achieves this by eliminating pathogenic microbes and lowering microbial numbers to prolong the quality of the beverage. Today, pasteurisation is used widely in the dairy industry and other food processing industries to achieve food preservation and food safety.

Unlike sterilization, pasteurization is not intended to kill all microorganisms in the food. Instead, it aims to reduce the number of viable pathogens so they are unlikely to cause disease (assuming the pasteurized product is stored as indicated and is consumed before its expiration date). Commercial-scale sterilization of food is not common because it adversely affects the taste and quality of the product. Certain foods, such as dairy products, may be superheated to ensure pathogenic microbes are destroyed.

Pasteurization Conditions

Minimum pasteurization requirements for milk products are shown in Table 1 below, and are based on regulations outlined in the Grade A Pasteurized Milk Ordinance (PMO). These conditions were determined to be the minimum processing conditions needed to kill Coxiella burnetii, the organism that causes Q fever in humans, which is the most heat resistant pathogen currently recognized in milk. Milk can be pasteurized using processing times and temperatures greater than the required minimums.

Pasteurization can be done as a batch or a continuous process. A vat pasteurizer consists of a temperature-controlled, closed vat. The milk is pumped into the vat, the milk is heated to the appropriate temperature and held at that temperature for the appropriate time and then cooled. The cooled milk is then pumped out of the vat to the rest of the processing line, for example to the bottling station or cheese vat. Batch pasteurization is still used in some smaller processing plants. The most common process used for fluid milk is the continuous process. The milk is pumped from the raw milk silo to a holding tank that feeds into the continuous pasteurization system. The milk continuously flows from the tank through a series of thin plates that heat up the milk to the appropriate temperature. The milk flow system is set up to make sure that the milk stays at the pasteurization temperature for the appropriate time before it flows through the cooling area of the pasteurizer. The cooled milk then flows to the rest of the processing line, for example to the bottling station. There are several options for temperatures and times available for continuous processing of refrigerated fluid milk. Although processing conditions are defined for temperatures above 200 °F, they are rarely used because they can impart an undesirable cooked flavor to milk.

Alcoholic Beverages

The process of heating wine for preservation purposes has been known in China since 1117, and was documented in Japan in the diary *Tamonin-nikki*, written by a series of monks between 1478 and 1618.

Much later, in 1768, an Italian priest and scientist Lazzaro Spallanzani proved experimentally that heat killed bacteria, and that they do not re-appear if the product is hermetically sealed. In 1795, a Parisian chef and confectioner named Nicolas Appert began experimenting with ways to preserve foodstuffs, succeeding with soups, vegetables, juices, dairy products, jellies, jams, and syrups. He placed the food in glass jars, sealed them with cork and sealing wax and placed them in boiling water. In that same year, the French military offered a cash prize of 12,000 francs for a new method to preserve food. After some 14 or 15 years of experimenting, Appert submitted his invention and won the prize in January 1810. Later that year, Appert published *L'Art de conserver les substances animales et végétales* (or *The Art of Preserving Animal and Vegetable Substances*). This was the first cookbook of its kind on modern food preservation methods.

La Maison Appert (English: The House of Appert), in the town of Massy, near Paris, became the first food-bottling factory in the world, preserving a variety of food in sealed bottles. Appert's method was to fill thick, large-mouthed glass bottles with produce of every description, ranging from beef and fowl to eggs, milk and prepared dishes. His greatest success for publicity was an entire sheep. He left air space at the top of the bottle, and the cork would then be sealed firmly in the jar by using a vise. The bottle was then wrapped in canvas to protect it, while it was dunked into boiling water and then boiled for as much time as Appert deemed appropriate for cooking the contents thoroughly. Appert patented his method, sometimes called in his honor *appertisation*.

Appert's method was so simple and workable that it quickly became widespread. In 1810, British inventor and merchant Peter Durand, also of French origin, patented his own method, but this time in a tin can, so creating the modern-day process of canning foods. In 1812, Englishmen Bryan Donkin and John Hall purchased both patents and began producing preserves. Just a decade later, Appert's method of canning had made its way to America. Tin can production was not common until the beginning of the 20th century, partly because a hammer and chisel were needed to open cans until the invention of a can opener by an inventor named Yates in 1855.

Appert's preservation by boiling involved heating the food to an unnecessarily high temperature, and for an unnecessarily long time, which could destroy some of the flavor of the preserved food.

A less aggressive method was developed by the French chemist Louis Pasteur during an 1864 summer holiday in Arbois. To remedy the frequent acidity of the local wines, he found out experimentally that it is sufficient to heat a young wine to only about 50–60 °C (122–140 °F) for a brief time to kill the microbes, and that the wine could subsequently be aged without sacrificing the final quality. In honour of Pasteur, the process became known as "pasteurization" Pasteurization was originally used as a way of preventing wine and beer from souring, and it would be many years before milk was pasteurized. In the United States in the 1870s, it was common for milk to contain substances intended to mask spoilage before milk was regulated.

Milk

Milk is an excellent medium for microbial growth, and when stored at ambient temperature bacteria and other pathogens soon proliferate.

400 lbs of milk in cheese vat

The US Centers for Disease Control (CDC) says improperly handled raw milk is responsible for nearly three times more hospitalizations than any other food-borne disease source, making it one of the world's most dangerous food products. Diseases prevented by pasteurization can include tuberculosis, brucellosis, diphtheria, scarlet fever, and Q-fever; it also kills the harmful bacteria *Salmonella*, *Listeria*, *Yersinia*, *Campylobacter*, *Staphylococcus aureus*, and *Escherichia coli* O157:H7, among others.

Pasteurization is the reason for milk's extended shelf life. High-temperature, short-time (HTST) pasteurized milk typically has a refrigerated shelf life of two to three weeks, whereas ultra-pasteurized milk can last much longer, sometimes two to three months. When ultra-heat treatment (UHT) is combined with sterile handling and container technology (such as aseptic packaging), it can even be stored unrefrigerated for up to 9 months.

History

Louis Pasteur's pasteurization experiment illustrates the fact that the spoilage of liquid was caused by particles in the air rather than the air itself. These experiments were important pieces of evidence supporting the idea of Germ Theory of Disease.

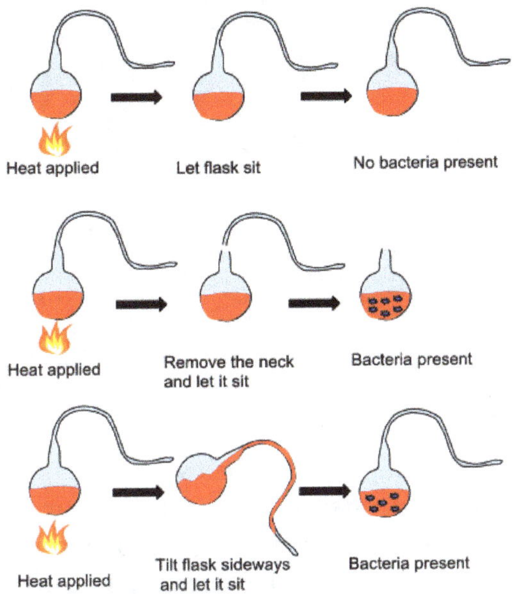

Before the widespread urban growth caused by industrialization, people kept dairy cows even in urban areas and the short time period between production and consumption minimized the disease risk of drinking raw milk. As urban densities increased and supply chains lengthened to the distance from country to city, raw milk (often days old) became recognised as a source of disease. For example, between 1912 and 1937 some 65,000 people died of tuberculosis contracted from consuming milk in England and Wales alone.

Developed countries adopted milk pasteurization to prevent such disease and loss of life, and as a result milk is now widely considered one of the safest foods. A traditional form of pasteurization by scalding and straining of cream to increase the keeping qualities of butter was practiced in England before 1773 and was introduced to Boston in the US by 1773, although it was not widely practiced in the United States for the next 20 years. It was still being referred to as a "new" process in American newspapers as late as 1802. Pasteurization of milk was suggested by Franz von Soxhlet in 1886. In the early 20th century, Milton Joseph Rosenau, established the standards (i.e. low temperature, slow heating at 60 °C (140 °F) for 20 minutes) for the pasteurization of milk, while at the United States Marine Hospital Service, notably in his publication of The Milk Question (1912). States in the U.S.A. began enacting mandatory dairy pasteurization laws with the first in 1947, and in 1973 the U.S. Federal Government required pasteurization of milk used in any interstate commerce.

Process

General overview of the pasteurization process. The milk starts at the left and enters the piping with functioning enzymes that, when heat treated, become denatured and stop the enzymes from functioning. This helps to stop pathogen growth by stopping the functionality of the cell. The cooling process helps stop the milk from undergoing the Maillard reaction and caramelization. The pasteurization process also has the ability to heat the cells to the point that they burst from pressure build up.

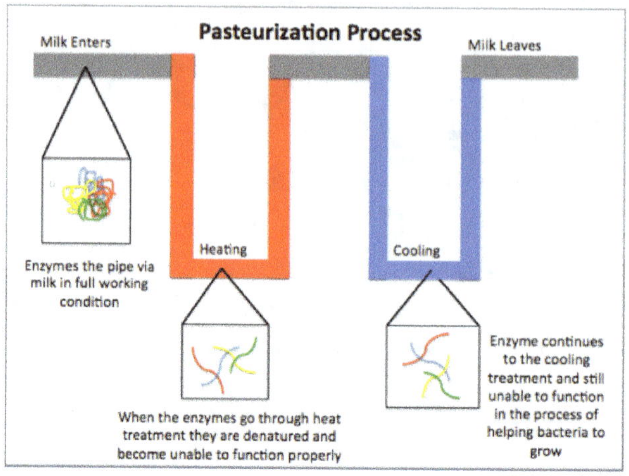

Older pasteurization methods used temperatures below boiling, since at very high temperatures, micelles of the milk protein casein irreversibly aggregate, or *curdle*. Newer methods use higher temperature, but shorten the time. Among the pasteurization methods listed below, the two main types of pasteurization used today are high-temperature, short-time (HTST, also known as "flash") and extended shelf life (ESL):

- HTST milk is forced between metal plates or through pipes heated on the outside by hot water, and the milk is heated to 72 °C (161 °F) for 15 seconds.[:8] Milk simply labeled "pasteurized" is usually treated with the HTST method.
- UHT, also known as ultra-heat-treating, processing holds the milk at a temperature of 140 °C (284 °F) for four seconds. During UHT processing milk is sterilized and not pasteurized. This process lets consumers store milk or juice for several months without refrigeration. The process is achieved by spraying the milk or juice through a nozzle into a chamber filled with high-temperature steam under pressure. After the temperature reaches 140 °C the fluid is cooled instantly in a vacuum chamber, and packed in a pre-sterilized airtight container. Milk labeled "ultra-pasteurized" or simply "UHT" has been treated with the UHT method.
- ESL milk has a microbial filtration step and lower temperatures than UHT milk. Since 2007, it is no longer a legal requirement in European countries (for example in Germany) to declare ESL milk as ultra-heated; consequently, it is now often labeled as "fresh milk" and just advertised as having an "extended shelf life," making it increasingly difficult to distinguish ESL milk from traditionally pasteurized fresh milk.
- A less conventional, but US FDA-legal, alternative (typically for home pasteurization) is to heat milk at 63 °C (145 °F) for 30 minutes.

Pasteurization methods are usually standardized and controlled by national food safety agencies (such as the USDA in the United States and the Food Standards Agency in the United Kingdom).

These agencies require that milk be HTST pasteurized to qualify for the pasteurized label. Dairy product standards differ, depending on fat content and intended usage. For example, pasteurization standards for cream differ from standards for fluid milk, and standards for pasteurizing cheese are designed to preserve the enzyme phosphatase, which aids cutting. In Canada, all milk produced at a processor and intended for consumption must be pasteurized, which legally requires that it be heated to at least 72 °C for at least 16 seconds, then cooling it to 4 °C to ensure any harmful bacteria are destroyed. The UK Dairy Products Hygiene Regulations 1995 requires that milk be heat treated for 15 seconds at 71.7 °C or other effective time/temperature combination.

Some older references point to one or multiple cycles of heating and cooling (to ambient temperature or below) as either a definition of pasteurisation or a general method thereof.

A process similar to pasteurization is thermization, which uses lower temperatures to kill bacteria in milk. It allows a milk product, such as cheese, to retain more of the original taste, but thermized foods are not considered pasteurized by food regulators.

Microwave Volumetric Heating

Microwave volumetric heating (MVH) is the newest available pasteurization technology. It uses microwaves to heat liquids, suspensions, or semi-solids in a continuous flow. Because MVH delivers energy evenly and deeply into the whole body of a flowing product, it allows for gentler and shorter heating, so that almost all heat-sensitive substances in the milk are preserved.

Efficiency

The HTST pasteurization standard was designed to achieve a five-log reduction, killing 99.999% of the number of viable micro-organisms in milk. This is considered adequate for destroying almost all yeasts, molds, and common spoilage bacteria and also to ensure adequate destruction of common pathogenic, heat-resistant organisms (including *Mycobacterium tuberculosis*, which causes tuberculosis, but not *Coxiella burnetii,* which causes Q fever). As a precaution, modern equipment tests and identifies bacteria in milk being processed. HTST pasteurization processes must be designed so the milk is heated evenly, and no part of the milk is subject to a shorter time or a lower temperature.

Even pasteurization without quality control can be effective, though this is generally not permitted for human consumption; a study of farms feeding calves on pasteurized waste milk using a mixture of pasteurization technologies (none of which were routinely monitored for performance) found the resulting pasteurized milk to meet safety requirements at least 92% of the time.

An effect of the heating of pasteurization is that some vitamin, mineral, and beneficial (or probiotic) bacteria is lost. Soluble calcium and phosphorus levels decrease by 5%, thiamine (vitamin B_1) and vitamin B_{12} (cobalamin) levels by 10%, and vitamin C levels by 20%. These losses are not significant nutritionally.

Verification

Direct microbiological techniques are the ultimate measurement of pathogen contamination, but these are costly and time-consuming (24–48 hours), which means that products are able to spoil by the time pasteurization is verified.

As a result of the unsuitability of microbiological techniques, milk pasteurization efficacy is typically monitored by checking for the presence of alkaline phosphatase, which is denatured by pasteurization. B. tuberculosis, the bacterium that requires the highest temperature to be killed of all milk pathogens is killed at ranges of temperature and time similar to those that denature alkaline phosphatase. For this reason, presence of alkaline phosphatase is an ideal indicator of pasteurization efficacy.

Phosphatase denaturing was originally monitored using a phenol-phosphate substrate. When hydrolysed by the enzyme these compounds liberate phenols, which were then reacted with dibromoquinonechlorimide to give a colour change, which itself was measured by checking absorption at 610 nm (spectrophotometry). Some of the phenols used were inherently coloured (phenolpthalein, nitrophenol) and were simply assayed unreacted. Spectrophotometric analysis is satisfactory but is of relatively low accuracy because many natural products are coloured. For this reason, modern systems (since 1990) use fluorometry which is able to detect much lower levels of raw milk contamination.

Unpasteurized Milk

According to the United States Centers for Disease Control between 1998 and 2011 79% of the dairy related outbreaks were due to raw milk or cheese products. They report 148 outbreaks, 2,384 illnesses (284 requiring hospitalizations) as well as 2 deaths due to raw milk or cheese products during the same time period.

Consumer Acceptance

Although pasteurization has been practiced for a long time, some consumers contend that they should have the right to buy and sell un-pasteurized milk if they want to.

Some consumers also point out that government-enforced pasteurization law has been used as a tool for large business to shut out competition from smaller producers. See the case of the FDA's shut down of Goodflow Juice in 2008.

Preservative

A preservative is a substance or a chemical that is added to products such as food, beverages, pharmaceutical drugs, paints, biological samples, cosmetics, wood, and many other products to prevent decomposition by microbial growth or by undesirable chemical changes. In general, preservation is implemented in two modes, chemical and physical. Chemical preservation entails adding chemical compounds to the product.Physical preservation entails processes such as refrigeration or drying. Preservative food additives reduce the risk of foodborne infections, decrease microbial spoilage, and preserve fresh attributes and nutritional quality. Some physical techniques for food preservation include dehydration, UV-C radiation, freeze-drying, and refrigeration. Chemical preservation and physical preservation techniques are sometimes combined.

Antimicrobial Additives

Antimicrobial preservatives prevent degradation by bacteria. This method is the most traditional and ancient type of preserving—ancient methods such as pickling and adding honey prevent microorganism growth by modifying the pH level. The most commonly used antimicrobial preservative is lactic acid. Common antimicrobial preservatives are presented in the table. Nitrates and nitrites are also antimicrobial. The detailed mechanism of these chemical compounds range from inhibiting growth of the bacteria to the inhibition of specific enzymes.

E number	chemical compound	comment
E201 – E203	benzoic acid, sodium benzoate	used in acidic foods such as jams, salad dressing, juices, pickles, carbonated drinks, soy sauce
E214 – E219	hydroxybenzoate and derivatives	stable at a broad pH range
E270	lactic acid	-
E249 – E250	nitrite	used in meats to prevent botulism toxin
E251 – E252	nitrate	used in meats
E280 – E283	propionic acid and sodium propionate	baked goods
E220 – E227	sulfur dioxide and sulfites	common for fruits
E200 – E203	sorbic acid and sodium sorbate	common for cheese, wine, baked goods

Antioxidants

The oxidation process spoils most food, especially those with a high fat content. Fats quickly turn rancid when exposed to oxygen. Antioxidants prevent or inhibit the oxidation process. The most common antioxidant additives are ascorbic acid (vitamin C) and ascorbates. Thus, antioxidants are commonly added to oils, cheese, and chips. Other antioxidants include the phenol derivatives BHA, BHT, TBHQ and propyl gallate. These agents suppress the formation of hydroperoxides. Other preservatives include ethanol and methylchloroisothiazolinone.

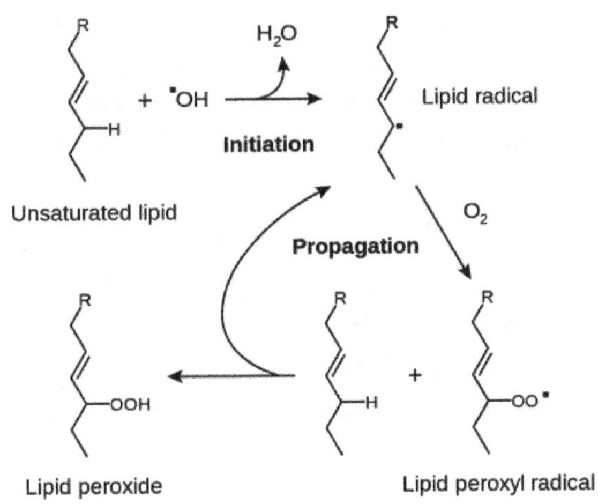

The free radical pathway for the first phase of the oxidative rancidification of fats. This process is slowed by antioxidants.

E number	chemical compound	comment
E300-304	ascorbic acid, sodium ascorbate	cheese, chips
E321	butylated hydroxytoluene, butylated hydroxyanisole	also used in food packaging
E310-312	gallic acid and sodium gallate	oxygen scavenger
E220 – E227	sulfur dioxide and sulfites	beverages, wine
E306 – E309	tocopherols	vitamin E activity

A variety of agents are added to sequester (deactivate) metal ions that otherwise catalyze the oxidation of fats. Common sequestering agents are disodium EDTA, citric acid (and citrates), tartaric acid, and lecithin.

Nonsynthetic Compounds for Food Preservation

Citric and ascorbic acids target enzymes that degrade fruits and vegetables, e.g., phenolase which turns surfaces of cut apples and potatoes brown. Ascorbic acid and tocopherol, which are vitamins, are common preservatives. Smoking entails exposing food to a variety of phenols, which are antioxidants. Natural preservatives include rosemary extract, hops, salt, sugar, vinegar, alcohol, diatomaceous earth and castor oil.

Traditional preservatives, such as sodium benzoate have raised health concerns in the past. Benzoate was shown in a study to cause hypersensitivity in some asthma sufferers. This has caused reexamination of natural preservatives which occur in vegetables.

History

Preservatives have been used since prehistoric times. Smoked meat for example has phenols and other chemicals that retard spoilage. The preservation of foods has evolved greatly over the centuries and has been instrumental in increasing food security. The use of preservatives other than traditional oils, salts, paints, etc. in food began in the late 19th century, but was not widespread until the 20th century.

The use of food preservatives varies greatly depending on the country. Many developing countries that do not have strong governments to regulate food additives face either harmful levels of preservatives in foods or a complete avoidance of foods that are considered unnatural or foreign. These countries have also proven useful in case studies surrounding chemical preservatives, as they have been only recently introduced. In urban slums of highly populated countries, the knowledge about contents of food tends to be extremely low, despite consumption of these imported foods.

Public Awareness of Food Preservation

Public awareness of food preservatives is uneven. Americans have a perception that food-borne illnesses happen more often in other countries. This may be true, but the occurrence of illnesses, hospitalizations, and deaths are still high. It is estimated by the Center for Disease Control (CDC) that each year there are 76 million illnesses, 325,000 hospitalizations, and 5,000 deaths linked to food-borne illness.

The increasing demand for ready-to-eat fresh food products has led to challenges for food distributors regarding the safety and quality of their foods. Artificial preservatives meet some of these challenges by preserving freshness for longer periods of time, but these preservatives can cause negative side-effects as well. Sodium nitrite is a preservative used in lunch meats, hams, sausages, hot dogs, and bacon to prevent botulism. It serves the important function of controlling the bacteria that cause botulism, but sodium nitrite can react with proteins, or during cooking at high heats, to form carcinogenic N-nitrosamines. It has also been linked to cancer in lab animals. The commonly used sodium benzoate has been found to extend the shelf life of bottled tomato paste to 40 weeks without loss of quality. However, it can form the carcinogen benzene when combined with vitamin C. Many food manufacturers have reformed their products to eliminate this combination, but a risk still exists. Consumption of sodium benzoate may also cause hyperactivity. For over 30 years, there has been a debate about whether or not preservatives and other food additives can cause hyperactivity. Studies have found that there may be increases in hyperactivity amongst children who consume artificial colorings and benzoate preservatives and who are already genetically predisposed to hyperactivity, but these studies were not entirely conclusive. Hyperactivity only increased moderately, and it was not determined if the preservatives, colorings, or a combination of the two were responsible for the increase.

Electroporation

Electroporation, or electropermeabilization, is a microbiology technique in which an electrical field is applied to cells in order to increase the permeability of the cell membrane, allowing chemicals, drugs, or DNA to be introduced into the cell. In microbiology, the process of electroporation is often used to transform bacteria, yeast, or plant protoplasts by introducing new coding DNA. If bacteria and plasmids are mixed together, the plasmids can be transferred into the bacteria after electroporation, though depending on what is being transferred cell-penetrating peptides or CellSqueeze could also be used. It works by passing several hundred volts across a distance of several millimeters are typically used in this process. Afterwards, the cells have to be handled carefully until they have had a chance to divide, producing new cells that contain reproduced plasmids. This process is approximately ten times more effective than chemical transformation.

Electroporation is also highly efficient for the introduction of foreign genes into tissue culture cells, especially mammalian cells. For example, it is used in the process of producing knockout mice, as well as in tumor treatment, gene therapy, and cell-based therapy. The process of introducing foreign DNA into eukaryotic cells is known as transfection. Electroporation is highly effective for transfecting cells in suspension using electroporation cuvettes. Electroporation has proven efficient for use on tissues in vivo, for in utero applications as well as in ovo transfection. Adherent cells can also be transfected using electroporation, providing researchers with an alternative to trypsinizing their cells prior to transfection.

Laboratory Practice

Electroporation is performed with electroporators, purpose-built appliances which create an electrostatic field in a cell solution. The cell suspension is pipetted into a glass or plastic cuvette which has two aluminum electrodes on its sides. For bacterial electroporation, typically a suspension of around 50 microliters is used. Prior to electroporation, this suspension of bacteria is mixed with the plasmid to be transformed. The mixture is pipetted into the cuvette, the voltage and capacitance are set, and the cuvette is inserted into the electroporator. The process requires direct contact between the electrodes and the suspension. Immediately after electroporation, one milliliter of liquid medium is added to the bacteria (in the cuvette or in an Eppendorf tube), and the tube is incubated at the bacteria's optimal temperature for an hour or more to allow recovery of the cells and expression of the plasmid, followed by bacterial culture on agar plates.

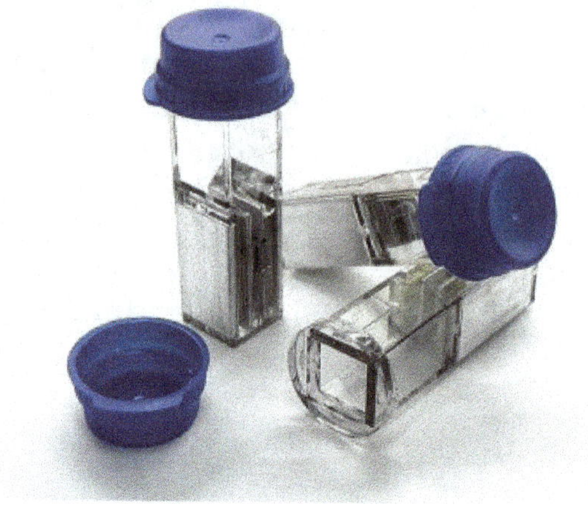

Cuvettes for electroporation. These are plastic with aluminium electrodes and a blue lid. They hold a maximum of 400 µl.

The success of the electroporation depends greatly on the purity of the plasmid solution, especially on its salt content. Solutions with high salt concentrations might cause an electrical discharge (known as arcing), which often reduces the viability of the bacteria. For a further detailed investigation of the process, more attention should be paid to the output impedance of the porator device and the input impedance of the cells suspension (e.g. salt content).

Since the cell membrane is not able to pass current (except in ion channels), it acts as an electrical capacitor. Subjecting membranes to a high-voltage electric field results in their temporary breakdown, resulting in pores that are large enough to allow macromolecules (such as DNA) to enter or leave the cell.

Medical Applications

The first research looking at how electroporation might be used on human cells was conducted by researchers at Eastern Virginia Medical School and Old Dominion University, and published in 2003.

The first successful treatment of malignant cutaneous tumors implanted in mice was completed in 2007 by a group of scientists who achieved complete tumor ablation in 12 out of 13 mice. They accomplished this by sending 80 pulses of 100 microseconds at 0.3 Hz with an electrical field magnitude of 2500 V/cm to treat the cutaneous tumors.

A higher voltage of electroporation was found in pigs to irreversibly destroy target cells within a narrow range while leaving neighboring cells unaffected, and thus represents a promising new treatment for cancer, heart disease and other disease states that require removal of tissue. Irreversible electroporation (IRE) has since proven effective in treating human cancer, with surgeons at Johns Hopkins and other institutions now using the technology to treat pancreatic cancer previously thought to be unresectable.

A recent technique called non-thermal irreversible electroporation (N-TIRE) has proven successful in treating many different types of tumors and other unwanted tissue. This procedure is done using small electrodes (about 1mm in diameter), placed either inside or surrounding the target tissue to apply short, repetitive bursts of electricity at a predetermined voltage and frequency. These bursts of electricity increase the resting transmembrane potential (TMP), so that nanopores form in the plasma membrane. When the electricity applied to the tissue is above the electric field threshold of the target tissue, the cells become permanently permeable from the formation of nanopores. As a result, the cells are unable to repair the damage and die due to a loss of homeostasis. N-TIRE is unique to other tumor ablation techniques in that it does not create thermal damage to the tissue around it.

Contrastingly, reversible electroporation occurs when the electricity applied with the electrodes is below the electric field threshold of the target tissue. Because the electricity applied is below the cells' threshold, it allows the cells to repair their phospholipid bilayer and continue on with their normal cell functions. Reversible electroporation is typically done with treatments that involve getting a drug or gene (or other molecule that is not normally permeable to the cell membrane) into the cell. Not all tissue has the same electric field threshold; therefore careful calculations need to be made prior to a treatment to ensure safety and efficacy.

One major advantage of using N-TIRE is that, when done correctly according to careful calculations, it only affects the target tissue. Proteins, the extracellular matrix, and critical structures such as blood vessels and nerves are all unaffected and left healthy by this treatment. This allows for a quicker recovery, and facilitates a more rapid replacement of dead tumor cells with healthy cells.

Before doing the procedure, scientists must carefully calculate exactly what needs to be done, and treat each patient on an individual case-by-case basis. To do this, imaging technology such as CT scans and MRI's are commonly used to create a 3D image of the tumor. From this information, they can approximate the volume of the tumor and decide on the best course of action including the insertion site of electrodes, the angle they are inserted in, the voltage needed, and more, using software technology. Often, a CT machine will be used to help with the placement of electrodes during the procedure, particularly when the electrodes are being used to treat tumors in the brain.

The entire procedure is very quick, typically taking about five minutes. The success rate of these procedures is high and is very promising for future treatment in humans. One disadvantage to using N-TIRE is that the electricity delivered from the electrodes can stimulate muscle cells to contract, which could have lethal consequences depending on the situation. Therefore, a paralytic agent must be used when performing the procedure. The paralytic agents that have been used in such research are successful; however, there is always some risk, albeit slight, when using anesthetics.

A more recent technique has been developed called high-frequency irreversible electroporation (H-FIRE). This technique uses electrodes to apply bipolar bursts of electricity at a high frequency, as opposed to unipolar bursts of electricity at a low frequency. This type of procedure has the same tumor ablation success as N-TIRE. However, it has one distinct advantage, H-FIRE does not cause muscle contraction in the patient and therefore there is no need for a paralytic agent.

Drug and Gene Delivery

Electroporation can also be used to help deliver drugs or genes into the cell by applying short and intense electric pulses that transiently permeabilize cell membrane, thus allowing transport of molecules otherwise not transported through a cellular membrane. This procedure is referred to as electrochemotherapy when the molecules to be transported are chemotherapeutic agents or gene electrotransfer when the molecule to be transported is DNA. Scientists from Karolinska Institutet and the University of Oxford use electroporation of exosomes to deliver siRNAs, antisense oligonucleotides, chemotherapeutic agents and proteins specifically to neurons after inject them systemically (in blood). Because these exosomes are able to cross the blood brain barrier this protocol could solve the issue of poor delivery of medications to the central nervous system and cure Alzheimer's, Parkinson's Disease and brain cancer among other diseases.

Physical Mechanism

Electroporation allows cellular introduction of large highly charged molecules such as DNA which would never passively diffuse across the hydrophobic bilayer core. This phenomenon indicates that the mechanism is the creation of nm-scale water-filled holes in the membrane. Although electroporation and dielectric breakdown both result from application of an electric field, the mechanisms involved are fundamentally different. In dielectric breakdown the barrier material is ionized, creating a conductive pathway. The material alteration is thus chemical in nature. In contrast, during electroporation the lipid molecules are not chemically altered but simply shift position, opening up a pore which acts as the conductive pathway through the bilayer as it is filled with water.

Electroporation is a dynamic phenomenon that depends on the local transmembrane voltage at each point on the cell membrane. It is generally accepted that for a given pulse duration and shape, a specific transmembrane voltage threshold exists for the manifestation of the electroporation phenomenon (from 0.5 V to 1 V). This leads to the definition of an electric field magnitude threshold for electroporation (E_{th}). That is, only the cells within areas where $E \geqq E_{th}$ are electroporated. If a second threshold (E_{ir}) is reached or surpassed, electroporation will compromise the viability of the cells, *i.e.*, irreversible electroporation (IRE).

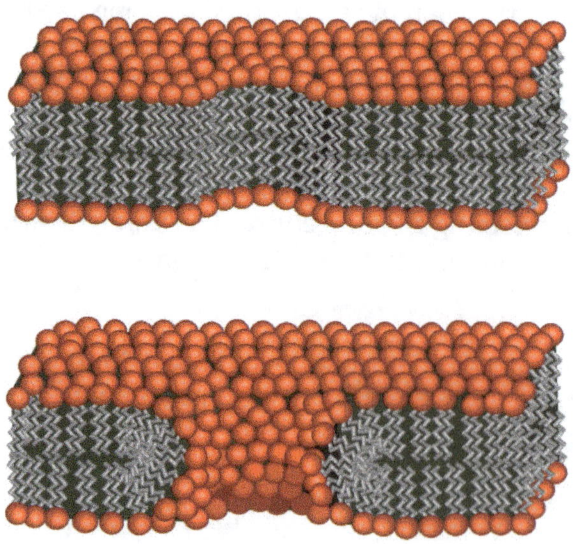

Schematic showing the theoretical arrangement of lipids in a hydrophobic pore (top) and a hydrophilic pore (bottom).

Electroporation is a multi-step process with several distinct phases. First, a short electrical pulse must be applied. Typical parameters would be 300–400 mV for < 1 ms across the membrane (note- the voltages used in cell experiments are typically much larger because they are being applied across large distances to the bulk solution so the resulting field across the actual membrane is only a small fraction of the applied bias). Upon application of this potential the membrane charges like a capacitor through the migration of ions from the surrounding solution. Once the critical field is achieved there is a rapid localized rearrangement in lipid morphology. The resulting structure is believed to be a "pre-pore" since it is not electrically conductive but leads rapidly to the creation of a conductive pore. Evidence for the existence of such pre-pores comes mostly from the "flickering" of pores, which suggests a transition between conductive and insulating states. It has been suggested that these pre-pores are small (~3 Å) hydrophobic defects. If this theory is correct, then the transition to a conductive state could be explained by a rearrangement at the pore edge, in which the lipid heads fold over to create a hydrophilic interface. Finally, these conductive pores can either heal, resealing the bilayer or expand, eventually rupturing it. The resultant fate depends on whether the critical defect size was exceeded which in turn depends on the applied field, local mechanical stress and bilayer edge energy.

Pascalization

Pascalization, bridgmanization, or high pressure processing (HPP), is a method of preserving and sterilizing food, in which a product is processed under very high pressure, leading to the inactivation of certain microorganisms and enzymes in the food. The technique was named after Blaise Pascal, a French scientist of the 17th century whose work included detailing the effects of pressure on fluids. During pascalization, more than 50,000 pounds per square inch (340 MPa, 3.4 kbar) may be applied for around fifteen minutes, leading to the inactivation of yeast, mold, and bacteria. Pascalization is also known as bridgmanization, named for physicist Percy Williams Bridgman.

Uses

Pascalization stops chemical activity caused by microorganisms that play a role in the deterioration of foods. The treatment occurs at low temperatures and does not include the use of food additives. From 1990, some juices, jellies, and jams have been preserved using pascalization in Japan. The technique is now used there to preserve fish and meats, salad dressing, rice cakes, and yogurts. HPP is now being used to preserve fruit and vegetable smoothies and other products such as meat for sale in the UK,. An early use of pascalization in the United States was to treat guacamole. It did not change the guacamole's taste, texture, or color, but the shelf life of the product increased to thirty days, from three days without the treatment. However, some treated foods still require cold storage because pascalization does not stop all enzyme activity caused by proteins, some of which affects shelf life.

History

Late 1800s

Experiments into the effects of pressure on microorganisms have been recorded as early as 1884, and successful experiments since 1897. In 1899, B. H. Hite was the first to conclusively demonstrate the inactivation of microorganisms by pressure. After he reported the effects of high pressure on microorganisms, and reports on the effects of pressure on foods quickly followed. Hite tried to prevent milk from spoiling, and his work showed that microorganisms can be deactivated by subjecting it to high pressure. He also mentioned some advantages of pressure-treating foods, such as the lack of antiseptics and no change in taste.

Hite said that, since 1897, a chemist at the West Virginia Agricultural Experimental Station had been studying the relationship between pressure and the preservation of meats, juices, and milk. Early experiments involved inserting a large screw into a cylinder and keeping it there for several days, but this did not have any effect in stopping the milk from spoiling. Later, a more powerful apparatus was able to subject the milk to higher pressures, and the treated milk was reported to stay sweeter for 24–60 hours longer than untreated milk. When 90 short tons (82 t) of pressure was applied to samples of milk for one hour, they stayed sweet for one week. Unfortunately, the device used to induce pressure was later damaged when researchers tried to test its effects on other products.

Experiments were also performed with anthrax, typhoid, and tuberculosis, which was a potential health risk for the researchers. Indeed, before the process was improved, one employee of the Experimental Station became ill with typhoid fever.

The process that Hite reported on was not feasible for widespread use and did not always completely sterilize the milk. While more extensive investigations followed, the original study into milk was largely discontinued due to concerns over its effectiveness. Hite mentioned "certain slow changes in the milk" related to "enzymes that the pressure could not destroy."

Early 1900s

Hite et al. released a more detailed report on pressure sterilization in 1914, which included the number of microorganisms that remained in a product after treatment. Experiments were conducted on various other foods, including fruits, fruit juices, and some vegetables. They were met with mixed success, similar to the results obtained from the earlier tests on milk. While some foods were preserved, others were not, possibly due to bacterial spores that had not been killed.

Hite's 1914 investigation led to other studies into the effect of pressure on microorganisms. In 1918, a study published by W. P. Larson et al. was intended to help advance vaccines. This report showed that bacterial spores were not always inactivated by pressure, while vegetative bacteria were usually killed. Larson et al.'s investigation also focused on the use of carbon dioxide, hydrogen, and nitrogen gas pressures. Carbon dioxide was found to be the most effective of the three at inactivating microorganisms.

Late 1900s–today

Around 1970, researchers renewed their efforts in studying bacterial spores after it was discovered that using moderate pressures was more effective than using higher pressures. These spores, which caused a lack of preservation in the earlier experiments, were inactivated faster by moderate pressure, but in a manner different from what occurred with vegetative microbes. When subjected to moderate pressures, bacterial spores germinate, and the resulting spores are easily killed using pressure, heat, or ionizing radiation. If the amount of initial pressure is increased, conditions are not ideal for germination, so the original spores must be killed instead. However, using moderate pressure does not always work, as some bacterial spores are more resistant to germination under pressure and a small portion of them will survive. A preservation method using both pressure and another treatment (such as heat) to kill spores has not yet been reliably achieved. Such a technique would allow for wider use of pressure on food and other potential advancements in food preservation.

Research into the effects of high pressures on microorganisms was largely focused on deep-sea organisms until the 1980s, when advancements in ceramic processing were made. This resulted in the production of machinery that allowed for processing foods at high pressures at a large scale, and generated some interest in the technique, especially in Japan. Although commercial products preserved by pascalization first emerged in 1990, the technology behind pascalization is still being perfected for widespread use. There is now higher demand for minimally processed products than in previous years, and products preserved by pascalization have seen commercial success despite being priced significantly higher than products treated with standard methods.

In the early 21st century, it was discovered that pascalization can separate the meat of shellfish from their shells. Lobsters, shrimp, crabs, etc. may be pascalized, and afterwards their raw meat will simply and easily slide whole right out of the cracked shell.

Process

In pascalization, food products are sealed and placed into a steel compartment containing a liquid, often water, and pumps are used to create pressure. The pumps may apply pressure constantly or intermittently. The application of high hydrostatic pressures (HHP) on a food product will kill many microorganisms, but the spores of some bacteria may need to be separately treated with acid to prevent their reproduction. Pascalization works especially well on acidic foods, such as yogurts and fruits, because pressure-tolerant spores are not able to live in environments with low pH levels. The treatment works equally well for both solid and liquid products.

During pascalization, the food's proteins are denatured, hydrogen bonds are fortified, and noncovalent bonds in the food are disrupted, while the product's main structure remains intact. Because pascalization is not heat-based, covalent bonds are not affected, causing no change in the food's taste. High hydrostatic pressure can affect muscle tissues by increasing the rate of lipid oxidation, which in turn leads to poor flavor and decreased health benefits.

Because hydrostatic pressure is able to act quickly and evenly on food, neither the size of a product's container nor its thickness play a role in the effectiveness of pascalization. There are several side effects of the process, including a slight increase in a product's sweetness, but pascalization does not greatly affect the nutritional value, taste, texture, and appearance. As a result, high pressure treatment of foods is regarded as a "natural" preservation method, as it does not use chemical preservatives.

Criticism

Anurag Sharma, a geochemist, James Scott, a microbiologist, and others at the Carnegie Institution of Washington directly observed microbial activity at pressures in excess of 1 gigapascal. The experiments were performed up to 1.6 GPa of pressure, which is more than 16,000 times normal air pressure, or about 14 times the pressure in the deepest ocean trench.

The experiment began by depositing an *Escherichia coli* and *Shewanella oneidensis* film in a Diamond Anvil Cell (DAC). The pressure was then raised to 1.6 GPa. When raised to this pressure and kept there for 30 hours, at least 1% of the bacteria survived. The experimenters then monitored formate metabolism using in-situ Raman spectroscopy and showed that formate metabolism continued in the bacterial sample.

Moreover, 1.6 GPa is such great pressure that during the experiment the DAC turned the solution into ice-IV, a room-temperature ice. When the bacteria broke down the formate in the ice, liquid pockets would form because of the chemical reaction.

There was some skepticism of this experiment. According to Art Yayanos, an oceanographer at the Scripps Institute of Oceanography, an organism should only be considered living if it can reproduce. Another issue with the DAC experiment is that when high pressures occur, there are usually high temperatures present as well, but in this experiment there were not. This experiment was performed at room-temperature. However, the intentional lack of high temperature in the experiments isolated the actual effects of pressure on life and results clearly indicated life to be largely pressure insensitive.

Newer results from independent research groups have confirmed Sharma et al. (2002). This is a significant step that reiterates the need for a new approach to the old problem of studying environmental extremes through experiments. There is practically no debate whether microbial life can survive pressures up to 600 MPa, which has been shown over the last decade or so to be valid through a number of scattered publications.

Consumer Acceptance of Pascalization

In the consumer studies of Hightech Europe consumers mentioned more positive than negative associations descriptions for this technology showing that these products are well accepted.

Food Irradiation

Cobalt-60 irradiation facility is used to test irradiation as a tool in the battle for food safety.

Food irradiation is the process of exposing foodstuffs to ionizing radiation. Ionizing radiation is energy that can be transmitted without direct contact (radiation) capable of freeing electrons from their atomic bonds (ionization) in the targeted food. This treatment is used to preserve food, reduce the risk of food borne illness, prevent the spread of invasive pests, and delay or eliminate sprouting or ripening. Irradiated food does not become radioactive. The radiation can be emitted by a radioactive substance or generated electrically.

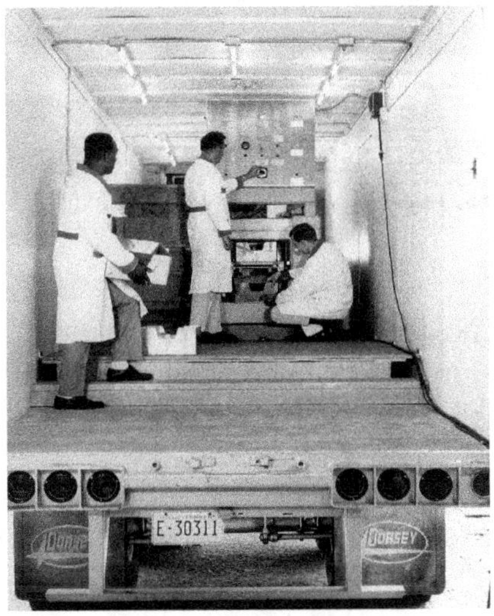

A portable, trailer-mounted food irradiation machine, circa 1968

Irradiation is also used for non-food applications, such as medical devices.

A large amount of independent research has confirmed irradiation to be safe. One family of chemicals is uniquely formed by irradiation, and this product is nontoxic. When irradiating food, all other chemicals occur in a lower or comparable frequency to other food processing techniques. Others criticize irradiation because of confusion with radioactive contamination or because of negative impressions of the nuclear industry.

Food irradiation is permitted by over 60 countries, with about 500,000 metric tons of food annually processed worldwide. The regulations that dictate how food is to be irradiated, as well as the food allowed to be irradiated, vary greatly from country to country. In Austria, Germany, and many other countries of the European Union only dried herbs, spices, and seasonings can be processed with irradiation and only at a specific dose, while in Brazil all foods are allowed at any dose.

Uses

Irradiation is used to reduce or eliminate the risk of food born illnesses. Depending on the dose, some or all of the pathogenic organism s, microorganisms, bacteria, and viruses present are destroyed, slowed down, or rendered incapable of reproduction. Some foods are irradiated at sufficient doses to ensure that the product is sterilized and does not add any spoilage or pathogenic microorganisms into the final product.

Irradiation is used reduce post harvest losses. It reduces the pathogens in the food and slows down the speed at which enzymes produced by the food can change the food and therefore slows the spoilage, ripening, and or sprouting of the food.

Irradiation cannot revert spoiled or over ripened food to a fresh state. If this food was processed by irradiation, spoilage would cease and ripening would slow down, yet the irradiation would not destroy the toxins or repair the texture, color, or taste of the food.

Food is irradiated to prevent the spread of foreign invasive species across national boundaries. The pests are sterilized or destroyed when the food is treated by low doses of irradiation. This also allows food to pass quickly through quarantine and avoid spoilage.

Public Perception and Impact

Irradiation has been approved by the FDA for over 50 years, but the only major growth area for the commercial sale of irradiated foods for human consumption is fruits and vegetables that are irradiated to kill insects for the purpose of quarantine. In the early 2000s in the US, irradiated meat was common at some grocery stores, but because of lack of consumer demand it is no longer common. Because consumer demand for irradiated food is low, reducing the spoilage between manufacture and consumer purchase and reducing the risk of food borne illness is currently not sufficient incentive for most manufactures to supplement their process with irradiation.

It is widely believed that consumer perception of foods treated with irradiation is more negative than those processed by other means, although some industry studies indicate the number of consumers concerned about the safety of irradiated food has decreased in the last 10 years to levels comparable to those of people concerned about food additives and preservatives. "These irradiated foods are not less safe than others," Dr. Tarantino said, "and the doses are effective in reducing the level of disease-causing micro-organisms." "People think the product is radioactive," said Harlan Clemmons, president of Sadex, a food irradiation company based in Sioux City, Iowa.

Some common concerns about food irradiation include the impact of irradiation on food chemistry, as well as the cumulative impacts of the irradiation process and the indirect effects of irradiation becoming a prevalent in the food handling process. Irradiation reduces the risk of infection and spoilage, does not make food radioactive, and the food is shown to be safe, but it does cause chemical reactions that alter the food and therefore alters the chemical makeup, nutritional content, and the sensory qualities of the food. Some of the potential secondary impacts of irradiation are hypothetical, while others are demonstrated. These effects include cumulative impacts to pathogens, people, and the environment due to the reduction of food quality, the transpiration and storage of radioactive goods, and destruction of pathogens. People are also concerned about the indirect effects of irradiation, such as the way irradiation changes the way we relate to food and how irradiation changes the food production and shipping industries. Because of these concerns and the increased cost of irradiated foods, there is not a widespread public demand for the irradiation of foods for human consumption.

Immediate Effects

The radiation source supplies energetic particles or waves (photons). As these waves/particles pass through a target material they collide with other particles. Around the sites of these collisions chemical bonds are broken, creating short lived radicals (e.g. the hydroxyl radical, the hydrogen atom and solvated electrons). These radicals cause further chemical changes by bonding with and or stripping particles from nearby molecules. When collisions damage DNA or RNA, effective reproduction becomes unlikely, also when collisions occur in cells, cell division is often suppressed.

Irradiation (within the accepted energy limits, as 10 MeV for electrons, 5 MeV for X-rays [US 7.5 MeV] and gamma rays from Cobalt-60) can not make food radioactive, but it does produce radiolytic products, and free radicals in the food. A few of these products are unique, but not considered dangerous.

Irradiation can also alter the nutritional content and flavor of foods, much like cooking. The scale of these chemical changes is not unique. Cooking, smoking, salting, and other less novel techniques, cause the food to be altered so drastically that its original nature is almost unrecognizable, and must be called by a different name. Storage of food also causes dramatic chemical changes, ones that eventually lead to deterioration and spoilage.

Misconceptions

A major concern is that irradiation might cause chemical changes that are harmful to the consumer. Several national expert groups and two international expert groups evaluated the available data and concluded that any food at any dose is wholesome and safe to consume as long as it remains palatable and maintains its technical properties (e.g. feel, texture, or color).

Irradiated food does not become radioactive, just as an object exposed to light does not start producing light. Radioactivity is the ability of a substance to emit high energy particles. When particles hit the target materials they may free other highly energetic particles. This ends shortly after the end of the exposure, much like objects stop reflecting light when the source is turned off and warm objects emit heat until they cool down but do not continue to produce their own heat. To modify a material so that it keeps emitting radiation (induce radiation) the atomic cores (nucleus) of the atoms in the target material must be modified.

It is impossible for food irradiators to induce radiation into a product. Irradiators emit electrons or photons and the radiation is intrinsically radiated at precisely known strengths (wavelengths for photons, and speeds for electrons). These radiated particles at these strengths can never be strong enough to modify the nucleus of the targeted atom in the food, regardless of how many particles hit the target material, and radioactivity can not be induced without modifying the nucleus.

Chemical Changes

Compounds known as free radicals form when food is irradiated. Most of these are oxidizers (i.e., accept electrons) and some react very strongly. According to the free-radical theory of aging excessive amounts of these free radicals can lead to cell injury and cell death, which may contribute to many diseases. However, this generally relates to the free radicals generated in the body, not the free radicals consumed by the individual, as much of these are destroyed in the digestive process.

When fatty acids are irradiated, a family of compounds called 2-alkylcyclobutanones (2-ACBs) are produced. These are thought to be unique radiolytic products. Most of the substances found in irradiated food are also found in food that has been subjected to other food processing treatments, and are therefore not unique. Furthermore, the quantities in which they occur in irradiated food are lower or similar to the quantities formed in heat treatments.

The radiation doses to cause toxic changes are much higher than the doses used to during irradiation, and taking into account the presence of 2-ACBs along with what is known of free radicals, these results lead to the conclusion that there is no significant risk from radiolytic products.

Food Quality

Because of the extent of the chemical reactions, changes to the foods quality after irradiation are inevitable. The nutritional contents, as well as the sensory qualities (taste, appearance, and texture) are impacted by irradiation. Irradiation changes the sensory qualities and reduces the content of some nutrients in foods. Irradiation also improves the shelf stability of some sensory qualities and nutrients. For these components more of the original taste, texture, appearance, and nutrients get to the final consumer. The changes in quality and nutrition vary greatly from food to food.

There has been low level gamma irradiation that has been attempted on arugula, spinach, cauliflower, ash gourd, bamboo shoots, coriander, parsley, and watercress. There has been limited information, however, regarding the physical, chemical and/or bioactive properties and the shelf life on these minimally processed vegetables.

Because the nutrational content changes after irradiation food advocacy groups consider labeling irradiated food raw as misleading. However, the degradation of vitamins caused by irradiation is similar or even less than the loss caused by other food preservation processes. Other processes like chilling, freezing, drying, and heating also result in some vitamin loss.

The changes in the flavor of fatty foods like meats, nuts and oils are sometimes noticeable, while the changes in lean products like fruits and vegetables are less so. Some studies by the irradiation industry show that for some properly treated fruits and vegetables irradiation is seen by consumers to improve the sensory qualities of the product compared to untreated fruits and vegetables.

Quality Impact on Minimally Processed Vegetables

Watercress (*Nasturtium Officinale*) is a rapidly growing aquatic or semi aquatic perennial plant. It contains health promoting phytochemicals endowed in therapeutic properties. Because chemical agents do not provide efficient microbial reductions, watercress has been tested with gamma irradiation treatment in order to improve both safety and the shelf life of the product. It is traditionally used on horticultural products to prevent sprouting and post-packaging contamination, delay post-harvest ripening, maturation and senescence.

In a Food Chemistry food journal, scientists studied the suitability of gamma irradiation of 1, 2, and 5 kGy for preserving quality parameters of the fresh cut watercress at around 4 degrees Celsius for 7 days. They determined that a 2 kGy dose of irradiation was the dose that contained most similar qualities to non-stored control samples, which is one of the goals of irradiation. 2 kGy preserved high levels of reducing sugars and favoured PUFA; while samples of the 5 kGy dose revealed high contents of sucrose and MUFA. Both cases the watercress samples obtained healthier fatty acids profiles. However, a 5kGy dose better preserved the antioxidant activity and total flavonoids.

Long Term Impacts

If the majority of food was irradiated at high enough levels to significantly decrease its nutritional content, there would be an increased risk to develop illnesses that are nutritional based if additional steps, such as changes in eating habits, where not taken to mitigate this. Furthermore, for at least 3 studies on cats, the consumption of irradiated food was associated with a loss of tissue in the myelin sheath, leading to reversible paralysis. Researchers suspect that reduced levels of vitamin A and high levels of free radicals may be the cause. This effect is thought to be specific to cats and has not been reproduced in any other animal. To produce these effects the cats were fed solely on food that was irradiated at a dose at least five times higher than the maximum allowable dose.

It may seem reasonable to assume that irradiating food might lead to radiation-tolerant strains, similar to the way that strains of bacteria have developed resistance to antibiotics. Bacteria develop a resistance to antibiotics after an individual uses antibiotics repeatedly. Much like pasteurization plants products that pass through irradiation plants are processed once, and are not processed and reprocessed. Cycles of heat treatment have been shown to produce heat tolerant bacteria, yet no problems have appeared so far in pasteurization plants. Furthermore, when the irradiation dose is chosen to target a specific species of microbe, it is calibrated to doses several times the value required to target the species. This ensures that the process randomly destroys all members of a target species. Therefore, the more irradiation tolerant members of the target species are not given any evolutionary advantage. Without evolutionary advantage selection does not occur. As to the irradiation process directly producing mutations that lead to more virulent, radiation resistant, strains the European Commission's Scientific Committee on Food found that there is no evidence, on the contrary, irradiation has been found to cause loss of virulence and infectivity as mutants are usually less competitive and less adapted."

Misconceptions

The argument is made that there is a lack of long-term studies, and therefore the safety of irradiated food is not scientifically proven in spite of the fact that hundreds of animal feeding studies of irradiated food, including multigenerational studies, have been performed since 1950. Endpoints investigated have included subchronic and chronic changes in metabolism, histopathology, and function of most systems; reproductive effects; growth; teratogenicity; and mutagenicity. A large number of studies have been performed; meta-studies have supported the safety of irradiated food.

The below experiments are cited by food irradiation opponents, but could be either not verified in later experiments, could not be clearly attributed to the radiation effect, or could be attributed to an inappropriate design of the experiment etc.

- India's National Institute of Nutrition (NIN) found an elevated rate of cells with more than one set of genes (Polyploidy) in humans and animals when fed wheat that was irradiated recently (within 12 weeks). Upon analysis scientist determined that the techniques used by NIN allowed for too much human error and statistical variation, therefore the results where unreliable. After multiple studies by independent agencies and scientists no correlation between polyploidy and irradiation of food could be found.

Indirect Effects of Irradiation

The indirect effects of irradiation are the concerns and benefits of irradiation that are related to how making food irradiation a common process will change the world, with emphasis on the system of food production.

If irradiation was to become common in the food handling process there would be a reduction of the prevalence of foodborne illness and potentially the eradication of specific pathogens. However, multiple studies suggest that an increased rate of pathogen growth may occur when irradiated food is cross-contaminated with a pathogen, as the competing spoilage organisms are no longer present. This being said, cross contamination itself becomes less prevalent with an increase in usage of irradiated foods.

The ability to remove bacterial contamination through post-processing by irradiation may reduce the fear of mishandling food which could cultivate a cavalier attitude toward hygiene and result in contaminants other than bacteria. However, concerns that the pasteurization of milk would lead to increased contamination of milk were prevalent when mandatory pasteurization was introduced, but these fears never materialized after adoption of this law. Therefore, it is unlikely for irradiation to cause an increase of illness due to nonbacteria-based contamination.

Treatment

Up to the point where the food is processed by irradiation, the food is processed in the same way as all other food. To treat the food, they are exposed to a radioactive source, for a set period of time to achieve a desired dose. Radiation may be emitted by a radioactive substance, or by X-ray and electron beam accelerators. Special precautions are taken to ensure the food stuffs never come in contact with the radioactive substances and that the personnel and the environment are protected from exposure radiation. Irradiation treatments are typically classified by dose (high, medium, and low), but are sometimes classified by the effects of the treatment (radappertization, radicidation and radurization). Food irradiation is sometimes referred to as "cold pasteurization" or "electronic pasteurization" because ionizing the food does not heat the food to high temperatures during the process, and the effect is similar to heat pasteurization. The term "cold pasteurization" is controversial because the term may be used to disguise the fact the food has been irradiated and pasteurization and irradiation are fundamentally different processes.

Treatment costs vary as a function of dose and facility usage. A pallet or tote is typically exposed for several minutes to hours depending on dose. Low-dose applications such as disinfestation of fruit range between US$0.01/lbs and US$0.08/lbs while higher-dose applications can cost as much as US$0.20/lbs.

Process

Typically, when the food is being irradiated, pallets of food are exposed to a source of radiation for a specific time. Dosimeters are embedded in the pallet (at various locations) of food to determine what dose was achieved. Most irradiated food is processed by gamma irradiation., however the usage of electron beam and X-ray is becoming more popular as well . Special precautions are taken because gamma rays are continuously emitted by the radioactive material. In most designs, to nullify the effects of radiation, the radioisotope is lowered into a water-filled storage pool, which absorbs the radiation but does not become radioactive. This allows pallets of the products to be added and removed from the irradiation chamber and other maintenance to be done. Sometimes movable shields are used to reduce radiation levels in areas of the irradiation chamber instead of submerging the source. For x ray and electron irradiation these precautions are not necessary as the source of the radiation can be switched off.

For x-ray, gamma ray and electron irradiation, shielding is required when the foods are being irradiated. This is done to protect workers and the environment outside of the chamber from radiation exposure. Typically permanent or movable shields are used. In some gamma irradiators the radioactive source is under water at all times, and the hermetically sealed product is lowered into the water. The water acts as the shield in this application. Because of the lower penetration depth of electron irradiation, treatment to entire industrial pallets or totes is not possible.

Dosimetry

The radiation absorbed dose is the amount energy absorbed per unit weight of the target material. Dose is used because, when the same substance is given the same dose, similar changes are observed in the target material. The SI unit for dose is grays (Gy or J/kg). Dosimeters are used to measure dose, and are small components that, when exposed to ionizing radiation, change measurable physical attributes to a degree that can be correlated to the dose received. Measuring dose (dosimetry) involves exposing one or more dosimeters along with the target material.

For purposes of legislation doses are divided into low (up to 1 kGy), medium (1 kGy to 10 kGy), and high-dose applications (above 10 kGy). High-dose applications are above those currently permitted in the US for commercial food items by the FDA and other regulators around the world. Though these doses are approved for non commercial applications, such as sterilizing frozen meat for NASA astronauts (doses of 44 kGy) and food for hospital patients.

Applications By Overall Average Dose		
	Application	**Dose (kGy)**
Low dose (up to 1 kGy)	Inhibit sprouting	0.03-0.15 kGy
	Delay fruit ripening	0.03-0.15 kGy
	Stop insect/parasite infestations	0.07-1.00 kGy
Medium dose (1 kGy to 10 kGy)	Delay spoilage of meat	1.50–3.00 kGy
	Reduce risk of pathogens in meat	3.00–7.00 kGy
	Increase sanitation of spices	10.00 kGy
High dose (above 10 kGy)	Sterilization of packaged meat	25.00-70.00 kGy
	Increase juice yield	
	Improve re-hydration	

Technology

Electron irradiation uses electrons accelerated in an electric field to a velocity close to the speed of light. Electrons have a charge, and therefore do not penetrate the product beyond a few centimeters, depending on product density.

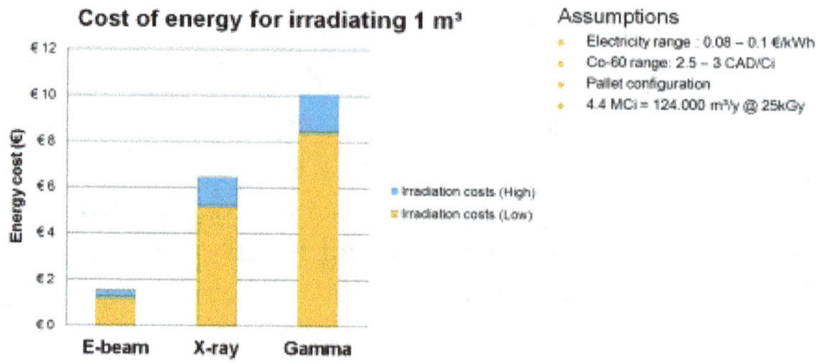

Efficiency illustration of the different radiation technologies (electron beam, X-ray, gamma rays)

Gamma irradiation involves exposing the target material to packets of light (photons) that are highly energetic (Gamma rays). A radioactive material (radioisotopes) is used as the source for the gamma rays. Gamma irradiation is the standard because the deeper penetration of the gamma rays enables administering treatment to entire industrial pallets or totes (reducing the need for material handling) and it is significantly less expensive than using a X-ray source. Generally cobalt-60 is used as a radioactive source for gamma irradiation. Cobalt-60 is bred from cobalt-59 using neutron irradiation in specifically designed nuclear reactors. In limited applications caesium-137, a less costly alternative recovered during the processing of spent nuclear fuel, is used as a radioactive source. Insufficient quantities are available for large scale commercial use. An incident where water-soluble caesium-137 leaked into the source storage pool requiring NRC intervention has led to near elimination of this radioisotope outside of military applications.

Irradiation by X-ray is similar to irradiation by gamma rays in that less energetic packets of light (X-rays) are used. X-rays are generated by colliding accelerated electrons with a dense material (this process is known as bremsstrahlung-conversion), and therefore do not necessitate the use of radioactive materials. X-rays ability to penetrate the target is similar to gamma irradiation. X-ray machines produce better dose uniformity than Gamma irradiation but require much more electricity as only as much as 12% of the input energy is converted into X-rays.

Cost

The cost of food irradiation is influenced by dose requirements, the food's tolerance of radiation, handling conditions, i.e., packaging and stacking requirements, construction costs, financing arrangements, and other variables particular to the situation. Irradiation is a capital-intensive technology requiring a substantial initial investment, ranging from $1 million to $5 million. In the case of large research or contract irradiation facilities, major capital costs include a radiation source, hardware (irradiator, totes and conveyors, control systems, and other auxiliary equipment), land (1 to 1.5 acres), radiation shield, and warehouse. Operating costs include salaries (for fixed and variable labor), utilities, maintenance, taxes/insurance, cobalt-60 replenishment, general utilities, and miscellaneous operating costs.

Standards & Regulations

The Codex Alimentarius represents the global standard for irradiation of food, in particular under the WTO-agreement. Member states are free to convert those standards into national regulations at their discretion, therefore regulations about irradiation differ from country to country.

The United Nations Food and Agricultural Organization (FAO) has passed a motion to commit member states to implement irradiation technology for their national phytosanitary programs; the General assembly of the International Atomic Energy Agency (IAEA) has urged wider use of the irradiation technology.

Labeling

The provisions of the Codex Alimentarius are that any "first generation" product must be labeled "irradiated" as any product derived directly from an irradiated raw material; for ingredients the provision is that even the last molecule of an irradiated ingredient must be listed with the ingredients even in cases where the unirradiated ingredient does not appear on the label. The RADURA-logo is optional; several countries use a graphical version that differs from the Codex-version. The suggested rules for labeling is published at CODEX-STAN – 1 (2005), and includes the usage of the Radura symbol for all products that contain irradiated foods. The Radura symbol is not a designator of quality. The amount of pathogens remaining is based upon dose and the original content and the dose applied can vary on a product by product basis.

The European Union follows the Codex's provision to label irradiated ingredients down to the last molecule of irradiated food. The European Community does not provide for the use of the Radura logo and relies exclusively on labeling by the appropriate phrases in the respective languages of the Member States. The European Union enforces its irradiation labeling laws by requiring its member countries to perform tests on a cross section of food items in the market-place and to report to the European Commission. The results are published annually in the OJ of the European Communities.

The US defines irradiated foods as foods in which the irradiation causes a material change in the food, or a material change in the consequences that may result from the use of the food. Therefore, food that is processed as an ingredient by a restaurant or food processor is exempt from the labeling requirement in the US. This definition is not consistent with the Codex Alimentarius. All irradiated foods must bear a slightly modified Radura symbol at the point of sale and use the term "irradiated" or a derivative there of, in conjunction with explicit language describing the change in the food or its conditions of use.

Food Safety

In 2003, the Codex Alimentarius removed any upper dose limit for food irradiation as well as clearances for specific foods, declaring that all are safe to irradiate. Countries such as Pakistan and Brazil have adopted the Codex without any reservation or restriction. Other countries, including New Zealand, Australia, Thailand, India, and Mexico, have permitted the irradiation of fresh fruits for fruit fly quarantine purposes, amongst others.

Standards that describe calibration and operation for radiation dosimetry, as well as procedures to relate the measured dose to the effects achieved and to report and document such results, are maintained by the American Society for Testing and Materials (ASTM international) and are also available as ISO/ASTM standards.

All of the rules involved in processing food are applied to all foods before they are irradiated.

United States

In the United States, each new food is approved separately with a guideline specifying a maximum dosage; in case of quarantine applications the minimum dose is regulated. Packaging materials containing the food processed by irradiation must also undergo approval. Food irradiation in the United States is primarily regulated by the FDA since it is considered a food additive. The United States Department of Agriculture (USDA) amends these rules for use with meat, poultry, and fresh fruit.

The United States Department of Agriculture (USDA) has approved the use of low-level irradiation as an alternative treatment to pesticides for fruits and vegetables that are considered hosts to a number of insect pests, including fruit flies and seed weevils. Under bilateral agreements that allows less-developed countries to earn income through food exports agreements are made to allow them to irradiate fruits and vegetables at low doses to kill insects, so that the food can avoid quarantine.

The U.S. Food and Drug Administration (FDA) and the USDA have approved irradiation of the following foods and purposes:

- Packaged refrigerated or frozen red meat — to control pathogens (E. Coli O157:H7 and Salmonella), and to extend shelf life.
- Packaged poultry — control pathogens (Salmonella and Camplylobacter).
- Fresh fruits, vegetables and grains — to control insects and inhibit growth, ripening and sprouting.
- Pork — to control trichinosis.
- Herbs, spices and vegetable seasonings — to control insects and microorganisms.
- Dry or dehydrated enzyme preparations — to control insects and microorganisms.
- White potatoes — to inhibit sprout development.
- Wheat and wheat flour — to control insects.
- Loose or bagged fresh iceberg lettuce and spinach

European Union

European law dictates that no foods other than dried aromatic herbs, spices and vegetable seasonings are permitted for the application of irradiation. However, any Member State is permitted to maintain previous clearances that are in categories that the EC's Scientific Committee on Food (SCF) had previously approved, or add clearance granted to other Member States. Presently, Belgium, Czech Republic, France, Italy, Netherlands, Poland, and the United Kingdom have adopted such provisions. Before individual items in an approved class can be added to the approved list, studies into the toxicology of each of such food and for each of the proposed dose ranges are requested. It also states that irradiation shall not be used "as a substitute for hygiene or health practices or good manufacturing or agricultural practice". These regulations only govern food irradiation in consumer products to allow irradiation to be used for patients requiring sterile diets.

Because of the Single Market of the EC any food, even if irradiated, must be allowed to be marketed in any other Member State even if a general ban of food irradiation prevails, under the condition that the food has been irradiated legally in the state of origin. Furthermore, imports into the EC are possible from third countries if the irradiation facility had been inspected and approved by the EC and the treatment is legal within the EC or some Member state.

Australia

Australia banned irradiated cat food after a national scare where cats suffered from paralyzation after eating a specific brand of highly irradiated catfood for an extended period of time. The suspected culprit was malnutrition from consuming food depleted of Vitamin A by the irradiation process. The incident was linked only to a single batch of one brand's product and no illness was linked to any of that brand's other irradiated batches of the same product or to any other brand of irradiated cat food. This, along with incomplete evidence indicating that the cat food was not sufficiently depleted of Vitamin A makes irradiation a less likely cause. Further research has been able to experimentally induce the paralyzation of cats by via Vitamin A deficiency by feeding highly irradiated food. For more details see the cumulative impacts of irradiation section.

Nuclear Safety & Security

Interlocks and safeguards are mandated to minimize this risk. There have been radiation related accidents, deaths, and injury at such facilities, many of them caused by operators overriding the safety related interlocks. In a radiation processing facility, radiation specific concerns are supervised by special authorities, while "Ordinary" occupational safety regulations are handled much like other businesses.

The safety of irradiation facilities is regulated by the United Nations International Atomic Energy Agency and monitored by the different national Nuclear Regulatory Commissions. The regulators enforce a safety culture that mandates that all incidents that occur are documented and thoroughly analyzed to determine the cause and improvement potential. Such incidents are studied by personnel at multiple facilities, and improvements are mandated to retrofit existing facilities and future design.

In the US the Nuclear Regulatory Commission (NRC) regulates the safety of the processing facility, and the United States Department of Transportation (DOT) regulates the safe transport of the radioactive sources.

Irradiated Food Supply

Authorities in some countries use tests that can detect the irradiation of food items to enforce labeling standards and to bolster consumer confidence. The European Union monitors the market to determine the quantity of irradiated foods, if irritated foods are labeled as irradiated, and if the irritation is performed at approved facilities.

Irradiation of fruits and vegetables to prevent the spread of pest and diseases across borders has been increasing globally. In 2010, 18446 tonnes of fruits and vegetables were irradiated in six countries for export quarantine control; the countries follow: Mexico (56.2%), United States (31.2%),

Thailand (5.18%), Vietnam (4.63%), Australia (2.69%), and India (0.05%). The three types of fruits irradiated the most were guava (49.7%), sweet potato(29.3%) and sweet lime (3.27%).

In total, 103 000 tonnes of food products were irradiated on mainland United States in 2010. The three types of foods irradiated the most were spices (77.7%), fruits and vegetables (14.6%) and meat and poultry (7.77%). 17 953 tonnes of irradiated fruits and vegetables were exported to the mainland United States. Mexico, the United States' state of Hawaii, Thailand, Vietnam and India export irradiated produce to the mainland U.S. Mexico, followed by the United States' state of Hawaii, is the largest exporter of irradiated produce to the mainland U.S.

In total, 6 876 tonnes of food products were irradiated in European Union countries in 2013; mainly in four member state countries: Belgium (49.4%), the Netherlands (24.4%), Spain (12.7%) and France (10.0%). The two types of foods irradiated the most were frog legs (46%), and dried herbs and spices (25%). There has been a decrease of 14% in the total quantity of products irradiated in the EU compared to the previous year 2012 (7 972 tonnes)

Timeline of the History of Food Irradiation

- 1895 Wilhelm Conrad Röntgen discovers X-rays ("bremsstrahlung", from German for radiation produced by deceleration)
- 1896 Antoine Henri Becquerel discovers natural radioactivity; Minck proposes the therapeutic use
- 1904 Samuel Prescott describes the bactericide effects Massachusetts Institute of Technology (MIT)
- 1906 Appleby & Banks: UK patent to use radioactive isotopes to irradiate particulate food in a flowing bed
- 1918 Gillett: U.S. Patent to use X-rays for the preservation of food
- 1921 Schwartz describes the elimination of Trichinella from food
- 1930 Wuest: French patent on food irradiation
- 1943 MIT becomes active in the field of food preservation for the U.S. Army
- 1951 U.S. Atomic Energy Commission begins to co-ordinate national research activities
- 1958 World first commercial food irradiation (spices) at Stuttgart, Germany
- 1970 Establishment of the International Food Irradiation Project (IFIP), headquarters at the Federal Research Centre for Food Preservation, Karlsruhe, Germany
- 1980 FAO/IAEA/WHO Joint Expert Committee on Food Irradiation recommends the clearance generally up to 10 kGy "overall average dose"
- 1981/1983 End of IFIP after reaching its goals
- 1983 Codex Alimentarius General Standard for Irradiated Foods: any food at a maximum

"overall average dose" of 10 kGy

- 1984 International Consultative Group on Food Irradiation (ICGFI) becomes the successor of IFIP
- 1998 The European Union's Scientific Committee on Food (SCF) voted "positive" on eight categories of irradiation applications
- 1997 FAO/IAEA/WHO Joint Study Group on High-Dose Irradiation recommends to lift any upper dose limit
- 1999 The European Union issues Directives 1999/2/EC (framework Directive) and 1999/3/EC (implementing Directive) limiting irradiation a positive list whose sole content is one of the eight categories approved by the SFC, but allowing the individual states to give clearances for any food previously approved by the SFC.
- 2000 Germany leads a veto on a measure to provide a final draft for the positive list.
- 2003 Codex Alimentarius General Standard for Irradiated Foods: no longer any upper dose limit
- 2003 The SCF adopts a "revised opinion" that recommends against the cancellation of the upper dose limit.
- 2004 ICGFI ends
- 2011 The successor to the SFC, European Food Safety Authority (EFSA), reexamines the SFC's list and makes further recommendations for inclusion.

Biopreservation

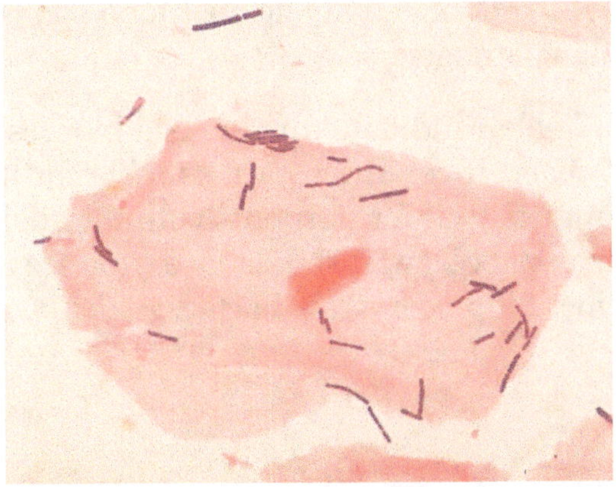

The small rods shown here are lactic acid bacteria which convert lactose and other sugars to lactic acid. The produts of their metabolism can have benign preservative effects.

Biopreservation

Biopreservation is the use of natural or controlled microbiota or antimicrobials as a way of preserving food and extending its shelf life. The biopreservation of food, especially utilizing lactic acid bacteria (LAB) that are inhibitory to food spoilage microbes, has been practiced since early ages, at first unconsciously but eventually with an increasingly robust scientific foundation. Beneficial bacteria or the fermentation products produced by these bacteria are used in biopreservation to control spoilage and render pathogens inactive in food. There are a various modes of action through which microorganisms can interfere with the growth of others such as organic acid production, resulting in a reduction of pH and the antimicrobial activity of the un-dissociated acid molecules, a wide variety of small inhibitory molecules including hydrogen peroxide, etc. It is a benign ecological approach which is gaining increasing attention.

Biopreservative Agents and Modes of Action

Lactic Acid Bacteria

Of special interest are lactic acid bacteria (LAB). Lactic acid bacteria have antagonistic properties which make them particularly useful as biopreservatives. When LABs compete for nutrients, their metabolites often include active antimicrobials such as lactic and acetic acid, hydrogen peroxide, and peptide bacteriocins. Some LABs produce the antimicrobial nisin which is a particularly effective preservative. These days LAB bacteriocins are used as an integral part of hurdle technology. Using them in combination with other preservative techniques can effectively control spoilage bacteria and other pathogens, and can inhibit the activities of a wide spectrum of organisms, including inherently resistant Gram-negative bacteria." Lactic acid bacteria and propionibacteria have been extensively studies for their efficacy against spoilage causing yeasts and molds in food spoilage.

Yeast

In addition to lactic acid bacteria, yeasts also have been reported to have a biopreservation effect due to their antagonistic activities relying on the competition for nutrients, production and tolerance of high concentrations of ethanol, as well as the synthesis of a large class of antimicrobial compounds exhibiting large spectrum of activity against food spoilage microorganisms, but also against plant, animal and human pathogen.

A bacterium/yeast that is a suitable candidate for use as a biopreservative does not necessarily have to ferment the food. However, if conditions are suitable for microbial growth, then a biopreservative bacterium will compete well for nutrients with the spoilage and pathogenic bacteria in the food. As a product of its metabolism, it should also produce acids and other antimicrobial agents, particularly bacteriocins. Biopreservative bacteria, such as lactic acid bacteria, must be harmless to humans.

Bacteriophages

Bacteriophages (Greek for 'bacteria eater'), or simply phages, are viruses which infect bacteria. The majority of all bacteriophages known exhibit a double-stranded DNA genome inside the virion capsid and belong to the order of tailed phages, Caudovirales. The tailed phages can be further separated into three families: Podoviridae, which are characterized by very short tails; Myoviridae, which exhibit longer, straight and contractile tails; and Siphoviridae, which can be identified due to their long and flexible tails. Another well studied group of phages with many applications, although minor in terms of species diversity, is represented by filamentous phages which exhibit a single stranded DNA genome decorated by a helical protein layer surrounding the DNA molecule. Bacteriophages are ubiquitously distributed in nature and can also be isolated from human or animal associated microflora. They outnumber their bacterial host species by a factor of ten representing the most abundant self-replicating entities on earth with an estimated 1031 phages in total. The idea of using phages against unwanted bacteria developed shortly after their discovery. With the improvements in organic chemistry during the 1950s, exploration and development of broad spectrum antibiotics displaced interest in bacteriophage research. Several laboratories have been testing suitability of bacteriophage isolates to control certain bacterial pathogens. Significant advancements in this research have been made at the Bacteriophage Institute in Tbilisi, Georgia, where phage therapy is routinely applied in medicine research field. Today treatment of antibiotic resistant bacteria is a challenging task. Recently, research on bacteriophages has gained additional momentum in light of the identification of antibiotic-resistant pathogens of infectious diseases, wherein the application of antibiotics is not effectively working, therefore research on the application of bacteriophages is being reviewed intensely. Bacteriophages have recently received a generally recognized as safe status based on their lack of toxicity and other detrimental effects to human health for application in meat products in USA.

Phage preparations specific for *L. monocytogenes, E. coli* O157:H7, and *S. enterica* serotypes have been commercialized and approved for application in foods or as part of surface decontamination protocols.

Meat Biopreservation

In meat processing, biopreservation has been extensively studied in fermented meat products and ready to eat meat products. The use of native or artificially-introduced microbial population to improve animal health and productivity, and/or to reduce pathogenic organisms, has been termed a probiotic or competitive enhancement approach. Competitive enhancement strategies that have been developed include competitive exclusion, addition of a microbial supplement (probiotic) that improves gastrointestinal health, and adding a limiting, non-host digestible nutrient (prebiotic) that provides an existing (or introduced) commensal microbial population a competitive advantage in the gastrointestinal tract. Each of these approaches utilizes the activities of the native microbial ecosystem against pathogens by capitalizing on the natural microbial competition. Generally speaking, competitive enhancement strategies offer a natural 'green' method to reduce pathogens in the gut of food animals.

Seafood Biopreservation

Fishery products are a source of wide variety of valuable nutrients such as proteins, vitamins, minerals, omega-3 fatty acids, taurine, etc. Fishery products, however, are also associated with human intoxication and infection. Approximately 10 to 20% of food-borne illnesses are attributed to fish consumption. Changing consumer demand has driven the appeal of traditional processes applied to seafood (e.g. salting, smoking and canning) lower compared to mild technologies involving lower salt content, lower cooking temperature and vacuum packing (VP)/modified atmosphere packing (MAP). These products, designed as lightly preserved fish products (LPFP), are usually produced from fresh seafood and further processing increases risk of cross contamination. These milder treatments are usually not sufficient to destroy microorganisms, and in some cases psychrotolerant pathogenic and spoilage bacteria can develop during the extended shelf-life of LPFP. Many of these products are also eaten raw, so minimizing the presence and preventing growth of microorganisms is essential for the food quality and safety. The microbial safety and stability of food are based on an application of preservative factors called hurdles. The delicate texture and flavor of seafood are very sensitive to the decontamination technologies such as cooking, and more recent mild technologies such as pulsed light, high pressure, ozone, and ultrasound. Chemical preservatives, which are not processes but ingredients, are out of favor with consumers due to natural preservatives demand. An alternative solution that is gaining more and more attention is biopreservation technology In fish processing, biopreservation is achieved by adding antimicrobials or by increasing the acidity of the fish muscle. Most bacteria stop multiplying when the pH is less than 4.5. Traditionally, acidity has been increased by fermentation, marination or by directly adding acetic, citric or lactic acid to food products. Other preservatives include nitrites, sulphites, sorbates, benzoates and essential oils. The main reason for less documented studies for application of protective microorganisms, bacteriophages or bacteriocins on seafood products for biopreservation compared to dairy or meat products is probably that the early stages of biopreservation have occurred mainly in fermented foodstuffs that are not so developed among seafood products. The selection of potential protective bacteria in seafood products is challenging due to the fact that they need adaptation to the seafood matrix (poor in sugar and their metabolic activities should not change the initial characteristics of the product, i.e. by acidification, and not induce spoilage that could lead to a sensory rejection. Among the microbiota identified in fresh or processed seafood, LAB remains the category that offers the highest potential for direct application as a bioprotective culture or for bacteriocin production.

Commercial Applications and Products

There has been successful implementation of various phage preparation around the globe. Various applications/delivery methods in food have been developed. Bacteriophages and their endolysins can be incorporated into food systems in several ways such as spraying, dipping or immobilization, singly or in combination with other hurdles. products The phage preparation LMP-1O2 has been subsequently commercialized as "ListShield" Intralyx, Inc. It has been shown to be effective against 170 different strains of "*L. monocytogenes*", reducing significantly (10 to 1000-fold) the Listeria contamination when sprayed onto ready-to-eat foods, without changing the food general composition, taste, odor or color. The Intralytix company has also commercialized phage-based antimicrobial preparations like SalmoFresh and SalmoLyse for controlling *S. enterica* SalmoFresh is prepared with a cocktail of naturally occurring lytic bacteriophages that selectively and specifically kill Salmonella, including strains belonging to the most common/highly pathogenic serotypes Typhimurium, Enteritidis, Heidelberg, Newport, Hadar, Kentucky and Thompson. According to the manufacturer, SalmoFresh is specifically designed for treating foods that are at high risk for "*Salmonella*" contamination. In particular, red meat and poultry can be treated prior to grinding for significant reductions in Salmonella contamination. SalmoLyse is a reformulated phage cocktail derived from SalmoFresh in which two of the six phages in the original cocktail have been replaced. Additional bacteriophage preparations have been formulated and referenced to be used to reduce the microbial load of animals prior to slaughter and are commercially available from Omnilytics such as the BacWash product line against Salmonella Omnilytics. Another commercial application has been developed, Listex_ P100 by Micreos in The Netherlands and was granted generally recognized as safe (GRAS) status by the FDA and USDA for use in all food products.

Another significant commercial bacteriophage application is ELICOSALI, a wide range of anti-*Salmonella* and "*E. coli*" phage cocktail, for treatment of agricultural products developed by Eliava Institute at Tbilisi, Republic of Georgia Eliava Institute.

Safety

Biopreservation judiciously exploits the antimicrobial potential of naturally occurring microorganisms in food and/or their metabolites with a long history of safe use. Bacteriocins, bacteriophages and bacteriophage-encoded enzymes fall in this theory. The long and traditional role of Lactic acid bacteria on food and feed fermentations is the main factor related to the use of bacteriocins in biopreservation. LAB and their bacteriocins have been consumed unintentionally for ages, laying down a long history of safe use. Their antimicrobial spectrum of inhibition, bactericidal mode of action, relative tolerance to processing conditions (pH, NaCl, heat treatments) and the lack of toxicity towards eukaryotic cells enforces their role as biopreservatives in food. The evaluation of any new antimicrobial actives is done in meat by USDA which relies on the GRAS assessment by FDA among other suitability data.

Food Storage

Yup'ik elevated food cache (*qulvarvik*), Hooper Bay, Alaska, 1929. Photograph by Edward S. Curtis

Food logistics is both a traditional domestic skill and is important industrially. The transport and storing of food as well as timely delivery to consumers is important in order to secure the procurement of food. Food is stored by almost every human society and by many animals. Storing of food has several main purposes:

U.S. Federal Emergency Management Agency (FEMA) food storage containers stacked on shipping pallets in Texas, 2008.

- Storage of harvested and processed plant and animal food products for distribution to consumers
- Enabling a better balanced diet throughout the year
- Reducing kitchen waste by preserving unused or uneaten food for later use
- Preserving pantry food, such as spices or dry ingredients like rice and flour, for eventual use in cooking
- Preparedness for catastrophes, emergencies and periods of food scarcity or famine
- Religious reasons (Example: LDS Church leaders instruct church members to store food)
- Protection from animals or theft

History of Food Storage

Domestic Food Storage

The safe storage of food for home use should strictly adhere to guidelines set out by reliable sources, such as the United States Department of Agriculture. These guidelines have been thoroughly researched by scientists to determine the best methods for reducing the real threat of food poisoning from unsafe food storage. It is also important to maintain proper kitchen hygiene, to reduce risks of bacteria or virus growth and food poisoning. The common food poisoning illnesses include Listeriosis, Mycotoxicosis, Salmonellosis, E. coli, Staphylococcal food poisoning and Botulism. There are many other organisms that can also cause food poisoning.

Tupperware kitchen storage containers designed for a variety of uses.

Plastic storage containers can be used to store food.

There are also safety guidelines available for the correct methods of home canning of food. For example, there are specific boiling times that apply depending upon whether pressure canning or waterbath canning is being used in the process. These safety guidelines are intended to reduce the growth of mold and bacteria and the threat of potentially-fatal food poisoning.

Food Storage Safety

Freezers and Thawing Food

Freezer temperature should be maintained below 0 °F. Food should never be thawed at room temperature, this increases the risk of bacterial and fungal growth and accordingly the risk of food poisoning. Once thawed, food should be used and never refrozen. Frozen food should be thawed using the following methods:

- Microwave oven
- During cooking
- In cold water (place food in watertight, plastic bag; change water every 30 minutes)
- In the refrigerator

Throw out foods that have been warmer than 40 °F for more than 2 hours. If there is any doubt at all about the length of time the food has been defrosted at room temperature, it should be thrown out. Freezing does not destroy microbes present in food. Freezing at 0 °F does inactivate microbes (bacteria, yeasts and molds). However, once food has been thawed, these microbes can again become active. Microbes in thawed food can multiply to levels that can lead to foodborne illness. Thawed food should be handled according to the same guidelines as perishable fresh food.

Food frozen at 0 °F and below is preserved indefinitely. However, the quality of the food will deteriorate if it is frozen over a lengthy period. The United States Department of Agriculture, Food Safety and Inspection Service publishes a chart showing the suggested freezer storage time for common foods.

Refrigeration

It is important to note that safe food storage using refrigeration requires adhering to temperature guidelines:

For safety, it is important to verify the temperature of the refrigerator. Refrigerators should be set to maintain a temperature of 40 °F or below. Some refrigerators have built-in thermometers to measure their internal temperature. For those refrigerators without this feature, keep an appliance thermometer in the refrigerator to monitor the temperature. This can be critical in the event of a power outage. When the power goes back on, if the refrigerator is still 40 °F, the food is safe. Foods held at temperatures above 40 °F for more than 2 hours should not be consumed. Appliance thermometers are specifically designed to provide accuracy at cold temperatures. Be sure refrigerator/freezer doors are closed tightly at all times. Don't open refrigerator/freezer doors more often than necessary and close them as soon as possible.

Storage Times for Refrigerated Food

The United States Department of Agriculture, Food Safety and Inspection Service publishes recommended storage times for refrigerated food.

Storing Oils and Fats

Oils and fats can begin to go rancid quickly when not stored safely. Rancid cooking oils and fats do not often smell rancid until well after they have spoiled. Oxygen, light and heat all contribute to cooking oils becoming rancid. The higher the level of polyunsaturated fat that an oil contains, the faster it spoils. The percentage of polyunsaturated fat in some common cooking oils is: safflower (74%); sunflower (66%); corn (60%); soybean (37%); peanut (32%); canola (29%); olive (8%).

To help preserve oils from rancidification, they should be refrigerated once opened. Opened, refrigerated cooking oils should be used within a few weeks, when some types begin to go rancid. Unopened oils can have a storage life of up to one year, but some types have a shorter shelf-life even when unopened (such as sesame and flaxseed).

Dry Storage of Foods

Vegetables

The guidelines vary for safe storage of vegetables under dry conditions (without refrigerating or freezing). This is because different vegetables have different characteristics, for example, tomatoes contain a lot of water, while root vegetables such as carrots and potatoes contain less. These factors, and many others, affect the amount of time that a vegetable can be kept in dry storage, as well as the temperature needed to preserve its usefulness. The following guideline shows the required dry storage conditions:

A large root cellar at the Oxon Hill Manor farm in Maryland

- Cool and dry: onion
- Cool and moist: root vegetable, potato, cabbage
- Warm and dry: winter squash, pumpkin, sweet potatoes, dried hot peppers

Many cultures have developed innovative ways of preserving vegetables so that they can be stored for several months between harvest seasons. Techniques include pickling, home canning, food dehydration, or storage in a root cellar.

Grain

Grain, which includes dry kitchen ingredients such as flour, rice, millet, couscous, cornmeal, and so on, can be stored in rigid sealed containers to prevent moisture contamination or insect or rodent infestation. For kitchen use, glass containers are the most traditional method. During the 20th century plastic containers were introduced for kitchen use. They are now sold in a vast variety of sizes and designs.

Metal cans are used (in the United States the smallest practical grain storage uses closed-top #10 metal cans). Storage in grain sacks is ineffective; mold and pests destroy a 25 kg cloth sack of grain in a year, even if stored off the ground in a dry area. On the ground or damp concrete, grain can spoil in as little as three days, and the grain might have to be dried before it can be milled. Food stored under unsuitable conditions should not be purchased or used because of risk of spoilage. To test whether grain is still good, it can be sprouted. If it sprouts, it is still good, but if not, it should not be eaten. It may take up to a week for grains to sprout. When in doubt about the safety of the food, throw it out as quickly as possible.

Spices and Herbs

Spices and herbs are today often sold prepackaged in a way that is convenient for pantry storage. The packaging has dual purposes of both storing and dispensing the spices or herbs. They are sold in small glass or plastic containers or resealable plastic packaging. When spices or herbs are homegrown or bought in bulk, they can be stored at home in glass or plastic containers. They can be stored for extended periods, in some cases for years. However, after 6 months to a year, spices and herbs will gradually lose their flavour as oils they contain will slowly evaporate during storage.

Spices and herbs can be preserved in vinegar for short periods of up to a month, creating a flavoured vinegar.

Alternative methods for preserving herbs include freezing in water or unsalted butter. Herbs can be chopped and added to water in an ice cube tray. After freezing, the ice cubes are emptied into a plastic freezer bag for storing in the freezer. Herbs also can be stirred into a bowl with unsalted butter, then spread on wax paper and rolled into a cylinder shape. The wax paper roll containing the butter and herbs is then stored in a freezer, and can be cut off in the desired amount for cooking. Using either of these techniques, the herbs should be used within a year.

Meat

Unpreserved meat has only a relatively short life in storage. Perishable meats should be refrigerated, frozen, dried promptly or cured. Storage of fresh meats is a complex discipline that affects the costs, storage life and eating quality of the meat, and the appropriate techniques vary with the kind of meat and the particular requirements. For example, dry ageing techniques are sometimes used to tenderize gourmet meats by hanging them in carefully controlled environments for up to 21 days, while game animals of various kinds may be hung after shooting. Details depend on personal tastes and local traditions. Modern techniques of preparing meat for storage vary with the type of meat and special requirements of tenderness, flavour, hygiene, and economy.

Semi-dried meats like salamis and country style hams are processed first with salt, smoke, sugar, acid, or other "cures" then hung in cool dry storage for extended periods, sometimes exceeding a year. Some of the materials added during the curing of meats serve to reduce the risks of food poisoning from anaerobic bacteria such as species of Clostridium that release botulinum toxin that can cause botulism. Typical ingredients of curing agents that inhibit anaerobic bacteria include nitrates. Such salts are dangerously poisonous in their own right and must be added in carefully controlled quantities and according to proper techniques. Their proper use has however saved many lives and much food spoilage.

Like the semi-dried meats, most salted, smoked, and simply-dried meats of different kinds that once were staples in particular regions, now are largely luxury snacks or garnishes; examples include jerky, biltong, and varieties of pemmican, but ham and bacon for instance, still are staples in many communities.

Food Rotation

Food rotation is important to preserve freshness. When food is rotated, the food that has been in storage the longest is used first. As food is used, new food is added to the pantry to replace it; the essential rationale is to use the oldest food as soon as possible so that nothing is in storage too long and becomes unsafe to eat. Labelling food with paper labels on the storage container, marking the date that the container is placed in storage, can make this practice simpler. The best way to rotate food storage is to prepare meals with stored food on a daily basis.

For Emergency Preparation

Guides for surviving emergency conditions in many parts of the world recommend maintaining a store of essential foods; typically water, cereals, oil, dried milk, and protein rich foods such as beans, lentils, tinned meat and fish. A food storage calculator can be used to help determine how much of these staple foods a person would need to store in order to sustain life for one full year. In addition to storing the basic food items many people choose to supplement their food storage with frozen or preserved garden-grown fruits and vegetables and freeze-dried or canned produce. An unvarying diet of staple foods prepared in the same manner can cause appetite exhaustion, leading to less caloric intake. Another benefit to having a basic supply of food storage in the home is for the potential cost savings. Costs of dry bulk foods (before preparation) are often considerably less than convenience and fresh foods purchased at local markets or supermarkets. There is a significant market in convenience foods for campers, such as dehydrated food products.

Commercial Food Logistics

Grain and beans are stored in tall grain elevators, almost always at a rail head near the point of production. The grain is shipped to a final user in hopper cars. In the former Soviet Union, where harvest was poorly controlled, grain was often irradiated at the point of production to suppress mold and insects. In the U.S., threshing and drying is performed in the field, and transport is nearly sterile and in large containers that effectively suppresses pest access, which eliminates the need for irradiation. At any given time, the U.S. usually has about two weeks worth of stored grains for the population.

Silos connected to a grain elevator on a farm in Israel.

Fresh fruits and vegetables are sometimes packed in plastic packages and cups for fresh premium markets, or placed in large plastic tubs for sauce and soup processors. Fruits and vegetables are usually refrigerated at the earliest possible moment, and even so have a shelf life of two weeks or less.

In the United States, livestock is usually transported live, slaughtered at a major distribution point, hung and transported for two days to a week in refrigerated rail cars, and then butchered and sold locally. Before refrigerated rail cars, meat had to be transported live, and this placed its cost so high that only farmers and the wealthy could afford it every day. In Europe much meat is transported live and slaughtered close to the point of sale. In much of Africa and Asia most meat is for local populations is raised, slaughtered and eaten locally, which is believed to be less stressful for the animals involved and minimizes meat storage needs. In Australia and New Zealand, where a large proportion of meat production is for export, meat enters the cold chain early, being stored in large freezer plants before being shipped overseas in freezer ships.

References

- Carlisle, Rodney (2004). Scientific American Inventions and Discoveries, p.357. John Wiley & Songs, Inc., new Jersey. ISBN 0-471-24410-4.
- Hwang, Andy; Huang, Lihan (31 January 2009). Ready-to-Eat Foods: Microbial Concerns and Control Measures. CRC Press. p. 88. ISBN 978-1-4200-6862-7. Retrieved 19 April 2011.
- Tortora, Gerard (2010). Microbiology: An Introduction, p.191. Pearson Benjamin Cummings, San Francisco. ISBN 0-321-55007-2.
- Whaites, Eric; Roderick Cawson (2002). Essentials of Dental Radiography and Radiology. Elsevier Health Sciences. pp. 15–20. ISBN 0-443-07027-X.
- Alzamora, Stella; Tapia, Maria Soledad; López-Malo, Aurelio (2000). Minimally Processed Fruits and Vegetables: Fundamental aspects and applications. Springer. p. 266. ISBN 978-0-8342-1672-3.
- Marotz, Lynn R. (2008). Health, Safety, and Nutrition for the Young Child. Wadsworth Publishing. p. 482. ISBN 978-1-4283-2070-3.
- Mitchell, Deborah (2004). Safe foods: the A-to-Z guide to the most wholesome foods for you and your family. Penguin. pp. Ch. 15. ISBN 978-1-101-21015-4.
- Editors of Creative Publishing. Dressing & Cooking Wild Game. Publisher: Creative Publishing international 1999 ISBN 978-0865731080

- "SCIENTIFIC STATUS SUMMARY Irradiation of Food". Institute of Food Technologists' Expert Panel on Food Safety and Nutrition in Food Technology. January 1998. Retrieved May 30, 2015.
- "Food Irradiation in Asia, the European Union, and the United States" (PDF). Japan Radioisotope Association. May 2013. Retrieved January 6, 2015.
- "Report from the Commission to the European Parliament and the Council on Food and Food Ingredients Treated with Ionising Radiation FOR THE YEAR 2013" (PDF). European Commission. February 25, 2015. Retrieved July 18, 2015.
- State Study Finds Calf Milk Pasteurization Effective, but Variable — Dairy — Penn State Extension". Extension. psu.edu. 2010-12-14. Retrieved 2014-03-19.

6

Food Marketing: An Integrated Study

Food marketing connects the producer to the consumer and involves multiple steps before the food becomes available to the buyer. This chapter explains in detail the concepts of food marketing, wholesale marketing of food and supermarkets. The reader is provided with thorough information on the types, trends, methods and representative bodies of each.

Food Marketing

Food marketing brings together the food producer and the consumer through a chain of marketing activities. The marketing of even a single food product can be a complicated process involving many producers and companies. For example, fifty-six companies are involved in making one can of chicken noodle soup. These businesses include not only chicken and vegetable processors but also the companies that transport the ingredients and those who print labels and manufacture cans. The food marketing system is the largest direct and indirect nongovernment employer in the United States.

Squashes and a pumpkin bought in the UK.

Food marketing is defined as a chain of marketing activities that take place within the food system between a food organisation and the consumer (Pomeranz & Adler, 2015). This has the potential to be a complicated procedure, as there are many processes that are used prior to the sale the food product. These include processing, wholesaling, retailing, food service and transport (Hawkes, 2007). Due to these many processes, a multitude of organisations have to be involved in the sale of one food product (Chandon, 2012). For example, approximately fifty-six organisations are involved in the making of one can of chicken noodle soup (Pomeranz & Adler, 2015). These organisations not only include the processors who make the ingredients for the product, but also involve the companies who manufacture the cans, print the labels and transport the product (Son, 2015). Therefore, on a global scale, the food marketing industry is one of the largest direct and indirect employers (Pomeranz & Adler, 2015).

Food marketing is the act of communicating to the consumer through a range of marketing techniques in order to add value to a food product and persuade the consumer to purchase. This includes all activities that occur in between the completion of a product through to the purchasing process of consumers. (Schaffner, Schroder & Earle 1997). Food marketing systems differ worldwide due to the level of development in the particular country, economically and technologically (Kaynak, 1999). Understanding and interpreting a particular countries food marketing techniques also requires taking into account the socio-economic, cultural, legal-political and technological environment of that country (Kaynak, 1999).

History

There are three historical phases of food marketing: the fragmentation phase (before 1870–1880), the unification phase (1880–1950), and the segmentation phase (1950 and later). In the fragmentation phase, the United States was divided into numerous geographic fragments because transporting food was expensive, leaving most production, distribution, and selling locally based. In the unification phase, distribution was made possible by railroads, coordination of sales forces was made possible by the telegraph and telephone, and product consistency was made possible by advances in manufacturing. This new distribution system was led by meat processors such as Armour and Swift in midwestern cities and by companies such as Heinz, Quaker Oats, Campbell Soup, and Coca-Cola, which sold their brands nationally. Advertising in print media and direct marketing through demonstrations at stores and public venues were among the prime marketing tools. The initial Crisco campaign, in 1911, was an example. In the segmentation phase (1950 and later) radio, television and internet advertising made it possible for a wider range of competing products to focus on different benefits and images and thus appeal to different demographic and psychographic markets. Distribution via the new national road system strengthened national brands.

History in The United States (U.S.)

The Fragmentation Phase (Pre-1870 - 1880)

At the time of this phase, the production, distribution and sale of food products was based locally, as the transportation of food in the U.S. was expensive (Murata, 2007). This was achieved by dividing the country into numerous segments, making it cheaper for food producers (Son, 2015).

The Unification Phase (1880 - 1950)

During this phase, the use of railroads made food distribution possible; telephones aided the coordination of sales; and advances in manufacturing technology made product consistency possible (Son, 2015). The prime marketing tools for this phase were advertising aspects, including print media, and direct marketing through store demonstrations (Murata, 2007). In the U.S., national organisations such as Coca-Cola, Campbell Soup and Heinz led this phase (Son, 2015).

The Segmentation Phase (1950 to Current)

Throughout this phase, organisations used television, internet and radio advertising in order to market its food products (Son, 2015). Through the use of these advertising methods, organisations were able to satisfy and understand particular geographic, demographic, psychographic and behavioural segments of the markets. This was achieved by displaying the different benefits and images of their food products (Murata, 2007).

Marketing Mix

The four components of food marketing are often called the "four Ps" of the marketing mix because they relate to product, price, promotion, and place. One reason food manufacturers receive the largest percentage of the retail food dollar is that they provide the most differentiating, value-added service. The money that manufacturers invest in developing, pricing, promotion, and placing their products helps differentiate a food product on the basis of both quality and brand-name recognition.

To understand how food marketing works as a whole, the "4 Ps" of the marketing mix must firstly be recognised. The 4Ps stand for product, price, promotion and place (Wongleedee, 2015). Organisations spend a large amount of revenue developing these aspects of the marketing mix in order to differentiate its food products, through quality and brand-name recognition (Pomeranz & Adler, 2015). Overall, the marketing mix can add value to a food organisation's product (Helm & Gritsch, 2014).

Product

In deciding what type of new food products a consumer would most prefer, a manufacturer can either try to develop a new food product or try to modify or extend an existing food. For example, a sweet, flavored yogurt drink would be a new product, but milk in a new flavor (such as chocolate strawberry) would be an extension of an existing product. There are three steps to both developing and extending: generate ideas, screen ideas for feasibility, and test ideas for appeal. Only after these steps will a food product make it to national market. Of one hundred new food product ideas that are considered, only six make it to a supermarket shelf.

The food industry faces numerous marketing decisions. Money can be invested in brand building (through advertising and other forms of promotion) to increase either quantity demanded or the price consumers are willing to pay for a product. Coca Cola, for example, spends a great deal of money both on perfecting its formula and on promoting the brand. This allows Coke to charge more for its product than can makers of regional and smaller brands. Manufacturers may be able to leverage their existing brand names by developing new product lines. For example, Heinz started out as a brand for pickles but branched out into ketchup. Some brand extensions may involve a risk of damage to the original brand if the quality is not good enough. Coca Cola, for example, refused to apply the Coke name to a diet drink back when artificial sweeteners had a significantly less attractive taste. Coke created Tab Cola, but only when aspartame (NutraSweet) was approved for use in soft drinks did Coca Cola come out with a Diet Coke. Manufacturers that have invested a great deal of money in brands may have developed a certain level of consumer brand loyalty—that is, a tendency for consumers to continue to buy a preferred brand even when an attractive offer is made by competitors. For loyalty to be present, it is not enough to merely observe that the consumer buys the same brand consistently. The consumer, to be brand loyal, must be able to actively resist promotional efforts by competitors. A brand loyal consumer will continue to buy the preferred brand even if a competing product is improved, offers a price promotion or premium, or receives preferred display space. Some consumers have multi-brand loyalty. Here, a consumer switches between a few preferred brands. The consumer may either alternate for variety or may, as a rule of thumb, buy whichever one of the preferred brands is on sale. This consumer, however, would not switch to other brands on sale. Brand loyalty is, of course, a matter of degree. Some consumers will not switch for a moderate discount, but would switch for a large one or will occasionally buy another brand for convenience or variety.

The product of the marketing mix refers to the goods and/or services that the organisation will offer to the consumer (Wongleedee, 2015). An organisation can achieve this by either creating a new food product, or by modifying or improving an existing food product (Pomeranz & Adler, 2015). For example, an organic almond yoghurt drink would be considered a new product, whereas a chocolate flavoured milk drink would be an extension of an existing product (Helm & Gritsch, 2014). The three steps to develop and extend a food product include generating ideas, analysing the ideas for feasibility and testing ideas for demand. Once these steps have successfully been completed, the food product can then be manufactured to the food market (Pomeranz & Adler, 2015).

Price

In profitably pricing the food, the manufacturer must keep in mind that the retailer adds approximately 50 percent to the price of a wholesale product. For example, a frozen food sold in a retail store for $4.50 generates an income of $3.00 for the manufacturer. This money has to pay for the cost of producing, packaging, shipping, storing, and selling the product.

Price encompasses the amount of money paid by the consumer in order to purchase the food product (Wongleedee, 2015). When pricing the food products, the manufacturer must bear in mind that the retailer will add a particular percentage to the price on the wholesale product. This percentage amount differs globally (Pomeranz & Adler, 2015). The percentage is used to pay for the cost of producing, packaging, shipping, storing and selling the food product. For example, the purchasing of a food product in a supermarket selling for $3.50 generates an income of $2.20 for the manufacturer (Wongleedee, 2015).

Promotion

Promoting a food to consumers is done out of store, in store, and on package. Advertisements on television and in magazines are attempts to persuade consumers to think favorably about a product, so that they go to the store to purchase the product. In addition to advertising, promotions can also include Sunday newspaper ads that offer coupons such as cents-off and buy-one-get-one-free offers.

Promotion of the marketing mix is defined as the actions used to communicate a food product's features and benefits; therefore, persuading the consumer to purchase the product (Wongleedee, 2015). There are multiple avenues used to promote a food product to consumers. Some examples are out-of-store, in-store and on the packaging (Helm & Gritsch, 2014). Food advertisements on television are used to attract the consumer, persuading them to go to the store and purchase that food product. Additionally, promotions in magazines and newspapers may offer coupons for food products (Frechette, 2015).

Place

Place refers to the distribution and warehousing efforts necessary to move a food from the manufacturer to a location where a consumer can buy it. It can also refer to where the product is located in a retail outlet (e.g., the end of an aisle; the top, bottom, or middle shelf; in a special display case, etc.).

The food marketing system in the United States is an amazingly flexible one. Consumer focus helps marketers anticipate the demands of consumers, and production focus helps them respond to changes in the market. The result is a system that meets and influences the ever-changing demands of consumers.

Place refers to the activities that organisations go about in order to make its food product available to its consumers. This encompasses the distribution necessary to move a food product from the manufacturer to a location where it can be purchased by the consumer (Wongleedee, 2015). Product location in a store is also a definition of place in the marketing mix. For example, a particular place in an aisle, a shelf or a display in a supermarket (Helm & Gritsch, 2014).

Food Marketing Segmentation

In order to market its food products, an organisation must first understand whether its product will satisfy the consumer's needs better than competitors do (Noori, 2015). In order to achieve this, an organisation must understand the four types of segmentation.

Geographic

An organisation must understand where it is marketing its food products to in a geographical sense (Schlanger & Maas, 2013). Clarifying this will help an organisation to grasp which food products will satisfy the needs of a particular consumer culture (Adeigbe, Baldwin, Gallion, Grier & Ramirez, 2015). For example, researching whether the consumer lives in America or Asia; or, whether the consumer lives in a city or a rural area (Schlanger & Maas, 2013). By understanding these aspects, the organisation will be able to identify that a consumers' food needs differ depending on where they live (Pomeranz & Adler, 2015).

Demographic

A food organisation must understand the demographic segment that it will be marketing towards (David, Schramm-Kelin, Rank & Wanger, 2015). Factors that must be considered are a consumer's age, gender, education, social class, income, religion and ethnicity (Quinn & Dibb, 2010). All of these aspects can impact whether the consumer will prefer one food product over another (Pomeranz & Adler, 2015).

Psychographic

A food organisation must understand its consumer psychographic. Factors such as the lifestyles, personalities, opinions, activities and interests of its potential consumers must considered (Schlanger & Maas, 2013). Identifying these aspects can help an organisation to improve its food products (Pomeranz & Adler, 2015).

Behaviour

A food organisation must understand how its consumers may behave towards a food product (David, Schramm-Kelin, Rank & Wanger, 2015). For example, information researching the benefits sought, frequency of food purchase, attitude towards the food product and nutritional knowledge of the product are all beneficial (Noori, 2015).

Issues in Food Marketing

Childhood Obesity:

It has been shown through Frechette's (2015) studies that in recent years, as children and teenagers have become more exposed to technological advances, they have become more susceptible to unhealthy food marketing commercial messages from food organisations (Cairns, Angus, Hastings & Caraher, 2013). Harris, Pomeranz, Lobstein, & Brownell (2009) suggest that food marketers have been using child-targeting marketing practices in order to persuade children to eat unhealthy or poorly suitable foods. Organisations achieve this through direct and indirect marketing tactics on television adverts, games, social media and food packaging (Freeman, Kelly, Baur, Chapman, Chapman, Gill & King, 2014). These tactics have an explicit effect on children's consumption patterns, diet-related health, nutritional knowledge, purchase behaviour and preferences (Vandevijvere & Swinburn, 2015). Overall, it has been found that food marketing is one of the leading contributors to an increase in childhood obesity (Schor & Ford, 2007). This is increasingly becoming a global issue (Chandon & Wansink, 2012).

In order to prevent the current unhealthy food marketing culture, Sacks, Mialon, Vandevijvere, Trevena, Snowdon, Crino & Swinburn (2015) believe that there are methods and policies that should be put into place by governments. Firstly, parents should be informed of the nutritional values of the foods that they are giving to their children (Grier, Mensinger, Huang, Kumanyika & Stettler, 2007). For example, an easy-to-read nutritional label on food packaging that provides the nutrient values and their definitions (Nikolova & Inman, 2015). This will create healthier food environments for families around the world (Sacks et al., 2015). Secondly, parents could restrict which advertisements that their children are exposed too (Newman & Oates, 2014). For example, parents could use ad-blocking applications or limit television watching time (Thaichon & Quach, 2016). Through the implementation of these strategies, governments can contribute in decreasing the rates of childhood obesity through food marketing (Hawkes, 2007).

Misleading Nutritional Information

In order to persuade consumers to buy food products, organisations may present misleading nutritional information on its food products (Schermel, Emrich, Arcand, Wong & L'abbé, 2013). It has become more prevalent that food organisations are not entirely transparent in regard to what is contained in its food products. For example, saturated fats, sodium and added sugars (Harris, LoDolce & Schwartz, 2015). Wording such as "less sugar", "fat free" and "all natural" lead consumers to believe that the foods they are consuming are healthy (Nikolova & Inman, 2015). Sacks et al. (2015) discusses that in order to prevent misleading food advertising, governments should implement policies regarding the placement of verified nutritional values on food packaging.

Criticism of Food Marketing

In recent years, food marketing has been criticised by professionals as being a major factor in contributing to childhood obesity. Nestle (2006) suggests that food marketing purposely targets children who are easily influenced at such a young age to eat high-sugar drinks and food with little nutritional benefit. The fact that areas of food marketing are linked to obesity is concerning, as well as how easy it is to market to children and affect their food choices. Television and print media still stand as the traditional communication channels for food marketing, but other popular approaches are also more prevalent these days. These include the Internet, toys, packaging, video games, blockbuster films, character licensing of children's toys and celebrity advertising (McGinnis, Appleton Gootman, & Kraak, 2006). The employment of these food marketing strategies are growing, and are said to be partly responsible for swelling rates of childhood obesity (Cartere, 2009).

Product placement in children's films and television shows gives food marketers more power to get children familiar with their brand and to directly interact with this market segment (Cartere, 2009). The power brands have through food marketing on television is significant because television audiences automatically are more enticed in an advertisement as it is playing in front of them, forming stronger predispositions for brands (Kline, 2010). Accusations come into play when this saturation happens as children are not equipped with adequate knowledge to make smart nutritional choices, and food marketing is therefore sometimes blamed for children's unhealthy lifestyles (Kline, 2010). Children are a fast expanding market segment, firstly because they yield influence over their parents buying, but also because they are future consumers themselves (Cartere, 2009). Food marketers capitalize on the fact most children trouble their parents for a product they have seen on television until they receive it, giving children high bargaining power. According to McGinnis et al. (2006), by the time children are two years old, the majority can identify brands in supermarkets and demand them by name. It has been argued that marketers also portray some foods to children as exclusive only to them, implying children know better about their nutrition than their parents. This has in turn seen trends of children disregarding parental advice and taking it upon themselves to pick and choose what they eat (Nestle, 2006).

Food marketers also use appealing packaging to attract children to their product through bright colours, including toys in schemes (McDonald's Happy Meals with a toy included is an example) and utilising famous television or film characters to spark interest (Cartere, 2009). In terms of packaging, brands will also change the size of products to entice children. Large companies have further been criticised of contributing to obesity through supplying schools with branded sponsorships and sports merchandise such as rugby balls that flaunt a company's logo (Cartere, 2009). Food marketers are criticised further than being responsible for child obesity rates, and are said to not have children's long-term physical wellbeing in mind when they aim to rapidly create brand name association among children (Frechette, 2015).

Purchasing Decisions in Food Marketing

Food marketing not only involves the marketing of products to consumers, but the reasons why consumers purchase these items and the factors influencing such choices (Sahay, Stough, Sohal, & Goyal, 2006). Demographics, values and attitudes, incentives, and price willingness to pay are all elements that drive buyer selection in the marketing of food.

Firstly, food marketers must be aware that the attitudes and values of their target market play a significant role in what they choose to buy. For example, in terms of the green brand market, consumers will first be environmentally conscious and therefore intend to buy such products. Once a product has been marketed to a consumer, they need to feel that they are contributing to the preservation of the environment to purchase something (Kapuge, 2016). Consumers that have concern for the environment tend to alter their behavior to be more environmentally friendly. Values and attitudes have been found to be the driving force of sustainable purchasing behavior (Kapuge, 2016). Marketers need to convey to their consumer market through information that some items have adverse effects on the environment. Brands need to also relay the values of food marketing to customers when communicating with them (Sahay et al., 2006).

In relation to knowledge and beliefs in variables affecting food marketing decisions, it has been suggested that someone's understanding of products helps anticipate their ecologically friendly actions (Nikolova & Inman, 2015). Because a persons awareness of information can cause them to make more informed or less informed decisions about products shows knowledge is a major factor in consumer buying choices. When strategists are food marketing, it is important to create binding relationships with potential customers through understanding their beliefs and awareness about the marketed product (Nikolova & Inman, 2015).

Demographics play a large role in determining the background of consumers and how food marketing operates. Such as age allows a brand to market its food towards certain age groups according to their wants. Education also comes into play because often education is associated with people buying better food (Sahay et al., 2006). Income alike sees if people have more disposable income to spend on good products. Gender further allows food marketers to target women especially because women have been found to do the majority of food shopping in families (Sahay et al., 2006).

Signage conveys a lot about a brand, for example what type of food they sell, the ingredients, where it was made and the cost of the products. Signage precisely communicates with the buyer and is an essential way to affect feelings and brain processes of a person (Remar, Campbell, & DiPietro, 2016). Findings show that certain digital signposting of a brand influences consumer perception and behavior towards a product, meaning food marketing strategies like this are very effective in brand positioning (Nikolova & Inman, 2015).

Lastly, consumer price willingness to pay is another tool used in food marketing to track how much a person is prepared to pay for a product. Marketers can take note of people's readiness to pay to see if they would indeed buy their product (Remar et al., 2016). Studies have shown that consumers are prepared to spend more on an item if it is environmentally friendly and portrays them as having sustainable behavior in society (Remar et al., 2016). Additionally, it has been explored that consumers will for example pay substantially more (almost 50%) for food produce that is said to be grown locally as opposed to grown in the wider country or imported (Remar et al., 2016). This is vital information when it comes to food marketing and helps companies make informed decisions about their food marketing strategies.

Food Marketing via Social Media and Its Influences

Social media platforms are an effective way for food marketers to enter or attract a market. 'Social media' refers to websites such as Facebook and Twitter, and apps such as Instagram and Facebook. These websites allow users to upload and share content such as information, pictures, opinions and recommendations with the touch of a button. Food companies use unique marketing techniques designed to connect with the consumer such as running competitions, holding free giveaways and creating interactive games and apps. (Freeman et al., 2014) The benefits of these techniques are distinctive to social media platforms, as they interact and engage with the consumer on a level that other traditional media platforms are not capable. (Freeman et al., 2014) The increase of social media platforms being used for food marketing are widely beneficial to food marketers, however the technological increase in society also allows consumers to block or delete advertisements while online and watching TV. AD Block and My Sky are prime examples of technological advances used by consumers to eliminate advertisements.

Supermarket Catalogues, Product Placement and their Involvement in Food Marketing

A study of supermarket catalogues distributed globally has also yielded information which supports the claim that food advertising is contributing to rising global obesity levels, especially in countries where the advertising is more prominent, consistent and persuasive. (Charlton, Kähkönen, Sacks & Cameron, 2015). The study concluded with the findings that supermarket catalogues have a major influence on the purchasing decisions and patterns of consumers. (Charlton et al., 2015) For example, countries with the most consistent, regular catalogues filled with unhealthy foods were among those testing highest for obesity as well. The study also noted that India and The Philippines were the only countries analysed that advertised a vast majority of healthy foods over unhealthy options. (Charlton et al., 2015).

Product placement is another food marketing technique that draws the consumer in and encourages them to purchase. Vending machines and pop-up sample booths are model examples of product placement techniques used by food marketers. Vending machines provide convenience to the consumer as they are usually stationed in areas with no larger food options nearby, such as university campuses and gyms. Consumers using vending machines will usually pay a slightly higher price for the item, and will usually not be concerned as the price difference accounts for the convenience provided to them.

Food Marketing Issues

There are no set worldwide food marketing laws or legislation, therefore countries have the option to adopt their own legislation with regards to food marketing standards. (Restrict Food Marketing, 2016). These standards are usually based on the values, culture and ethics of the particular country (Kaynak, 1999). Many countries worldwide have created laws in order to limit food marketing towards children with the goal of reducing rising obesity levels. (Restrict Food Marketing, 2016). Obesity has been proven to have increased significantly with a link to the increase of food marketing and advertising in society. (Dibb & Lobstein, 2005).

Following a 'call out' from the Worldwide Health Organisation in 2006, many countries adopted to change marketing laws and legislation in order to protect children from persuasive advertisements directly targeted at them. (Center for science in the public interest, 2007). These advertisements are strategically designed with special techniques in order to attract the children's attention.

The Chilean Government has taken some of the most drastic steps in 2012 by approving the 'Law of Nutritional Composition of Food and Advertising'. (Restrict Food Marketing, 2016). This law gives strict guidelines for food marketers, allowing no high in saturated fat, sodium and sugar products to be advertised to under 14 year olds or through television or websites with an audience or more than 20% children. (Restrict Food Marketing, 2016). Iran has also taken similar actions, having banned the advertisement of all soft drinks since 2004. Iran's Ministry of Health and Medical Education has also prepared a document in 2015 containing 24 items in which they hope to prohibit the advertising of in future. (Restrict Food Marketing, 2016). Dibb and Lobstein acknowledge that traditional media outlets only account for approximately 20% of a food marketer's budget, therefore children are still highly exposed to influential advertising in shopping malls and grocery stores, vending machines, sponsored toys, contests etc. (Dibb & Lobstein, 2005).

In addition to Iran and Chile's more extensive legislation's, over 30 countries worldwide have all adopted some form of legislation to protect children and reduce their exposure to advertisements. Some of these countries include Australia, Europe, Canada, Malaysia and Korea. (Center for science in the public interest, 2007). An example of the type of clauses in the legislation's include prohibiting the use of certain marketing techniques such as using cartoon characters, which can be used to access a child's mind and gain their attention. (Centre for science in the public interest, 2007). Using physiological techniques such as this can be ethically wrong as well as illegal in some countries.

Advertising to children at a young age is a well-established food marketing technique designed to encourage brand preference and holds many ethical dilemmas. Previous studies have concluded that children can recognize and mentally picture brand logos at the age of just six months old, and will verbally request brands at the age of 3 years. (As reported by The Campaign for a Commercial-Free Childhood). Marketing professor James McNeal acknowledged "The Drool Factor" – a study which recognized the fact that babies naturally stare down at their bibs while drooling to see where their drool lands. (Collins, 2013). Customizing baby bibs with brand logos has become an effective way for food marketers to imprint their brand into the child's lifestyle, targeting them at a vulnerable young age resulting in brand recognition from the child. (Collins, 2013) As a result, when the child is older they will continue to reciprocate warm, fond feelings towards the brand when encountering it in society. This is seen as unethical in many countries and is why organisations such as 'The Campaign for a Commercial-free Childhood' exist and what they aim to reduce.

Wholesale Marketing of Food

The consumption and production of marketed food are spatially separated. Production is primarily in rural areas while consumption is in urban areas. Agricultural marketing is the process that overcomes this separation, allowing produce to be moved from an area of surplus to one of need. Food reaches the consumer by a complex network, involving production, assembly, sorting, packing, reassembly, distribution and retail stages. In developing countries the linkage between the producer and the retailer is still usually provided by assembly and wholesale markets, where wholesale marketing takes place using a variety of transaction methods. Recent years have seen an expansion of wholesale marketing in E. European and former CIS countries. On the other hand, the growth of supermarkets in many regions has seen the development of direct marketing and a reduced role for wholesale systems.

In the fruits and vegetables sector of the Rungis International Market, France.

Types of Wholesale Market

Secondary wholesale markets are generally found only in developing countries these days. They are located in district or regional cities and take the bulk of their produce from rural assembly markets located in production areas, where the transactions are small scale and usually take place between farmers and traders. The distinction between rural assembly markets and secondary wholesale markets is that secondary wholesale markets are in permanent operation (rather than being seasonal in nature or dealing in specialized produce), larger volumes of produce are traded than at the rural assembly markets and specialized functions may be present, such as commission agents and brokers.Wholesaling include all the activity involved in selling goods and services to those buying for resale or business use .Firms engaged primarily in wholesaling are called wholesalers.

Wholesale vegetable market in Munich.

Wholesalers buy mostly from producers and sell mostly to retailers, industrial consumers, and other wholesalers.

Types of wholesalers:

1. Merchant wholesalers
2. Agents
3. Brokers

1) Merchant wholesalers are the largest single group of wholesalers, accounting for roughly 50 percent of all wholesaling. Merchant wholesalers include two broad types: full-service wholesalers and limited-service wholesalers .

2) Brokers and agents differ from merchant wholesalers in two ways: They do not take title to goods and they perform few function like merchant wholesalers, Like merchant wholesalers, they generally specialize by product line or customer type. A broker brings buyers and sellers together and assists in negotiation. Agents represent buyers or sellers on a more permanent basis.

Terminal wholesale markets are located in major metropolitan areas, where produce is finally channelled to consumers through trade between wholesalers and retailers, caterers, etc. Produce may also be assembled for export. In some countries, such as India and China, terminal markets also supply other parts of the country. For example, New Delhi serves as a distribution centre to the south of India for apples grown in the Himalayan foothills. The problems of terminal wholesale markets are usually ones of congestion caused by an unsuitable location or by an inappropriate mixture of wholesale and retail functions. Traditionally, wholesale markets were built adjacent to city centres, located at a focal point of the inter-city transport facilities and close to the main retailing areas. Population growth, changes in urban land-use patterns and the development of modern transport systems have all influenced the suitability and functionality of existing sites.

Transaction Methods

Wholesaling facilitates the economic function of buying and selling (usually termed as "price formation") by allowing the forces of supply and demand to converge to establish a single price for a commodity. The wholesaler may also perform storage and warehousing functions, as well as allowing economies of scale to be obtained in the transportation of produce from farm to market. The people involved in wholesaling can act simply as merchants, buying and selling produce, can be brokers dealing in orders rather than goods, or can be commission agents. Prices are normally established by negotiation but some wholesale markets use the auction system. The auction is compulsory in all wholesale markets in big cities in Japan. A comparison of the Japanese system with that in Brisbane, Australia, found that prices moved more frequently and with a greater degree of volatility in Japan than those at Brisbane.

Wholesale vegetable market in India.

Wholesaling Trends

Wholesale markets develop in a number of stages. They start as general markets, then become more specialized by trading in specific types of product. A later stage is to transact only graded and well packaged produce. A recent trend in Western Europe and the USA is for large retailers to by-pass the wholesale market system. Direct links are created between producers and supermarket chains, often by means of contract farming arrangements or through the use of preferred suppliers. Following the collapse of the iron curtain a large number of markets were developed in Eastern and Central Europe in the 1990s and early 2000s. Examples include markets in Warsaw, Gdansk, Budapest, and Bucharest. In other parts of the world, new markets have been built in Amman, Cairo and Mumbai, among many recent developments. Very few new wholesale markets have been built in western countries in the last decades, although old markets have been relocated to new sites (e.g. the new Covent Garden Market relocated to Nine Elms in London and Rome's new wholesale market, relocated from the centre to the east of the city). Those that already exist have tended to also attract warehouses for integrated food distribution, changing their role to "food centers" (in the USA) and including other non-fresh food products. Wholesale markets still have a role in the marketing of horticultural produce but the traditional fresh meat and fish wholesale markets, particularly those dealing with wholesale live produce, are generally being closed down in major urban centres. In developing countries, changes in work patterns, particularly the employment of women, and the impact of technological innovations in post-harvest handling, food processing and storage, including the use of domestic refrigerators, tends to encourage the development of one-stop shopping at supermarkets, often on a once-a-week basis. The challenge for wholesale markets in such a trading environment is to retain turnover, both by providing new services to supermarkets and by developing services to the non-supermarket trade and the growing hotel and catering sectors.

Haikou wholesale vegetable market, China.

Representative Bodies

The World Union of Wholesale Markets (WUWM) aims to represent wholesale markets internationally, and promote the exchange of information between them for mutual benefit.

Supermarket

A supermarket, a large form of the traditional grocery store, is a self-service shop offering a wide variety of food and household products, organized into aisles. It is larger and has a wider selection than a traditional grocery store, but is smaller and more limited in the range of merchandise than a hypermarket or big-box market.

Produce section in a Brazilian supermarket

The supermarket typically comprises meat, fresh produce, dairy, and baked goods aisles, along with shelf space reserved for canned and packaged goods as well as for various non-food items such as kitchenware, household cleaners, pharmacy products and pet supplies. Some supermarkets also sell a variety of other household products that are consumed regularly, such as new condoms (where permitted), medicine, and clothes, and some stores sell a much wider range of non-food products: DVDs, sporting equipment, board games, and seasonal items (e.g., Christmas wrapping paper in December).

The traditional supermarket occupies a large amount of floor space, usually on a single level. It is usually situated near a residential area in order to be convenient to consumers. The basic appeal is the availability of a broad selection of goods under a single roof, at relatively low prices. Other advantages include ease of parking and frequently the convenience of shopping hours that extend into the evening or even 24 hours of day. Supermarkets usually allocate large budgets to advertising, typically through newspapers. They also present elaborate in-shop displays of products. The shops are usually part of corporate chains that own or control (sometimes by franchise) other supermarkets located nearby—even transnationally—thus increasing opportunities for economies of scale.

Supermarkets typically are supplied by the distribution centres of their parent companies, usually in the largest city in the area. Supermarkets usually offer products at relatively low prices by using their buying power to buy goods from manufacturers at lower prices than smaller stores can. They also minimise financing costs by paying for goods at least 30 days after receipt and some extract credit terms of 90 days or more from vendors. Certain products (typically staple foods such as bread, milk and sugar) are very occasionally sold as loss leaders, that is, with negative profit margins so as to attract shoppers to their store. There is some debate as to the effectiveness of this tactic. To maintain a profit, supermarkets make up for the lower margins by a higher overall volume of sales, and with the sale of higher-margin items bought by the intended higher volume of shoppers. Customers usually shop by placing their selected merchandise into shopping carts (trolleys) or baskets (self-service) and pay for the merchandise at the check-out. At present, many supermarket chains are attempting to further reduce labor costs by shifting to self-service check-out machines, where a single employee can oversee a group of four or five machines at once, assisting multiple customers at a time.

A larger full-service supermarket combined with a department store is sometimes known as a hypermarket. Other services offered at some supermarkets may include those of banks, cafés, childcare centres/creches, Insurance(and other financial services), Mobile Phone services, photo processing, video rentals, pharmacies and/or petrol stations.

History

In the early days of retailing, all products generally were fetched by an assistant from shelves behind the merchant's counter while customers waited in front of the counter and indicated the items they wanted. Also, most foods and merchandise did not come in individually wrapped consumer-sized packages, so an assistant had to measure out and wrap the precise amount desired by the consumer. This also offered opportunities for social interaction: many regarded this style of shopping as "a social occasion" and would often "pause for conversations with the staff or other customers." These practices were by nature very labor-intensive and therefore also quite expensive. The shopping process was slow, as the number of customers who could be attended to at one time was limited by the number of staff employed in the store.

A supermarket in Sweden in 1941

Shopping for groceries also often involved trips to multiple specialty shops, such as a greengrocer, butcher, bakery, fishmonger and dry goods store, in addition to a general store, while milk was delivered by a milkman.

The concept of an inexpensive food market relying on large economies of scale was developed by Vincent Astor. He founded the Astor Market in 1915, investing $750,000 ($18 million in 2015 currency) of his fortune into a 165' by 125' corner of 95th and Broadway, Manhattan, creating, in effect, an open air mini-mall that sold meat, fruit, produce and flowers. The expectation was that customers would come from great distances ("miles around"), but in the end even attracting people from ten blocks away was difficult, and the market folded in 1917.

The concept of a self-service grocery store was developed by entrepreneur Clarence Saunders and his Piggly Wiggly stores. His first store opened in 1916. Saunders was awarded a number of patents for the ideas he incorporated into his stores. The stores were a financial success and Saunders began to offer franchises. The Great Atlantic & Pacific Tea Company, which was established in 1859, was another successful early grocery store chain in Canada and the United States, and became common in North American cities in the 1920s. The general trend in since then has been to stock shelves at night so that customers, the following day, can obtain their own goods and bring them to the front of the store to pay for them. Although there is a higher risk of shoplifting, the costs of appropriate security measures ideally will be outweighed by reduced labor costs.

Consumers shopping for produce and fruit.

Early self-service grocery stores did not sell fresh meats or produce. Combination stores that sold perishable items were developed in the 1920s.

Historically, there was debate about the origin of the supermarket, with King Kullen and Ralphs of California having strong claims. Other contenders included Weingarten's Big Food Markets and Henke & Pillot. To end the debate, the Food Marketing Institute in conjunction with the Smithsonian Institution and with funding from H.J. Heinz, researched the issue. It defined the attributes of a supermarket as "self-service, separate product departments, discount pricing, marketing and volume selling."

It has been determined that the first true supermarket in the United States was opened by a former Kroger employee, Michael J. Cullen, on 4 August 1930, inside a 6,000-square-foot (560 m^2) former garage in Jamaica, Queens in New York City. The store, King Kullen, operated under the slogan "Pile it high. Sell it low." At the time of Cullen's death in 1936, there were seventeen King Kullen stores in operation. Although Saunders had brought the world self-service, uniform stores and nationwide marketing, Cullen built on this idea by adding separate food departments, selling large volumes of food at discount prices and adding a parking lot.

Other established American grocery chains in the 1930s, such as Kroger and Safeway at first resisted Cullen's idea, but eventually were forced to build their own supermarkets as the economy sank into the Great Depression, while consumers were becoming price-sensitive at a level never experienced before. Kroger took the idea one step further and pioneered the first supermarket surrounded on all four sides by a parking lot.

As larger chain supermarkets began to dominate the market in the USA, able to supply consumers with the desired lower prices as opposed to the smaller "mom and pop" stands with considerable more overhead costs, the backlash of this infrastructure alteration was seen through numerous anti-chain campaigns. The idea of "monopsony," proposed by Cambridge economist Joan Robinson in 1933, that a single buyer could out-power the market of multiple sellers, became a strong anti-chain rhetorical device. With public backlash came political pressure to even the playing field for smaller vendors without the luxuries of economies of scale. In 1936, the Robinson-Patman Act was implemented as a way of preventing such larger chains from using this buying power to reap advantages over smaller, although the act was not well enforced and did not have much impact in the prevention of larger chains overtaking power in the markets.

A Safeway advertisement from the 1950s

Supermarkets proliferated across Canada and the United States with the growth of automobile ownership and suburban development after World War II. Most North American supermarkets are located in suburban strip shopping centers as an anchor store along. They are generally regional rather than national in their company branding. Kroger is perhaps the most nationally oriented supermarket chain in the United States but it has preserved most of its regional brands, including Ralphs, City Market, King Soopers, Fry's, Smith's, and QFC.

Pyatorochka supermarket in Moscow, Russia

In Canada, the largest such company is Loblaw, which operates stores under a variety of banners targeted to different segments and regions, including Fortinos, Zehrs, No Frills, the Real Canadian Superstore, and Loblaws, the foundation of the company. Sobeys is Canada's second largest supermarket with locations across the country, operating under many banners (Sobeys IGA in Quebec). Québec's first supermarket opened in 1934 in Montréal, under the banner Steinberg's.

In the United Kingdom, self-service shopping took longer to become established. Even in 1947, there were just ten self-service shops in the country. In 1951, ex-US Navy sailor Patrick Galvani, son-in-law of Express Dairies chairman, made a pitch to the board to open a chain of supermarkets across the country. The UK's first supermarket under the new Premier Supermarkets brand opened in Streatham, South London, taking ten times as much per week as the average British general store of the time. Other chains caught on, and after Galvani lost out to Tesco's Jack Cohen in 1960 to buy the 212 Irwin's chain, the sector underwent a large amount of consolidation, resulting in 'the big four' dominant UK of today: Tesco, Asda (owned by Wal-Mart), Sainsbury's and Morrisons.

In the 1950s, supermarkets frequently issued trading stamps as incentives to customers. Today, most chains issue store-specific "membership cards," "club cards," or "loyalty cards". These typically enable the card holder to receive special members-only discounts on certain items when the credit card-like device is scanned at check-out. Sales of selected data generated by clubcards is becoming a significant revenue stream for some supermarkets.

Traditional supermarkets in many countries face intense competition from discounters such as Wal-Mart, and Tesco in the UK, which typically are non-union and operate with better buying power. Other competition exists from warehouse clubs such as Costco that offer savings to customers buying in bulk quantities. Superstores, such as those operated by Wal-Mart and Asda, often offer a wide range of goods and services in addition to foods. The proliferation of such warehouse and superstores has contributed to the continuing disappearance of smaller, local grocery stores; increased dependence on the automobile; suburban sprawl because of the necessity for large floorspace and increased vehicular traffic. For example, in 2009 51% of Wal-Mart's $251 Billion domestic sales were recorded from grocery goods. Some critics consider the chains' common practice of selling loss leaders to be anti-competitive. They are also wary of the negotiating power that large, often multinational have with suppliers around the world.

Growth in Developing Countries

There has been a rapid transformation of the food sector in developing countries, beginning in the 1990s. This applies particularly to Latin America, South-East Asia, India, China and South Africa. However, growth is being witnessed in nearly all countries. With growth, has come considerable competition and some amount of consolidation. The growth has been driven by increasing affluence and the rise of a middle class; the entry of women into the workforce; with a consequent incentive to seek out easy-to-prepare foods; the growth in the use of refrigerators, making it possible to shop weekly instead of daily; and the growth in car ownership, facilitating journeys to distant stores and purchases of large quantities of goods. The opportunities presented by this potential have encouraged several European companies to invest in these markets (mainly in Asia) and American companies to invest in Latin America and China. Local companies also entered the market. Initial development of supermarkets has now been followed by hypermarket growth. In addition there were investments by companies such as Makro and Metro in large-scale Cash-and-Carry operations.

While the growth in sales of processed foods in these countries has been much more rapid than the growth in fresh food sales, the imperative nature of supermarkets to achieve economies of scale in purchasing, means that the expansion of supermarkets in these countries has important repercussions for small farmers, particularly those growing perishable crops. New supply chains have developed involving cluster formation; development of specialized wholesalers; leading farmers organizing supply; and farmer associations or cooperatives. In some cases supermarkets have organized their own procurement from small farmers; in others wholesale markets have adapted to meet supermarket needs.

Typical Supermarket Merchandise

Larger supermarkets in North America and in Europe typically sell a great number of items among many brands, sizes and varieties, including:

Inside an Asda supermarket in Keighley, West Yorkshire

Sainsbury's supermarket front end

- Alcoholic beverages (as state, provincial or local laws allow)
- Baby foods and baby-care products such as disposable diapers
- Breads and bakery products (many stores may have a bakery on site that offers specialty and dessert items)
- Books, newspapers, and magazines, including supermarket tabloids
- Bulk dried foods such as legumes, flour, rice, etc. (typically available for self-service)
- Canned goods and dried cereals
- CDs, audio cassettes, DVDs, and videos (including video rentals)
- Cigarettes and other tobacco products (as country, state, provincial or local laws allow)

- Confections and candies
- Cosmetics
- Dairy products and eggs
- Delicatessen foods (ready-to-eat)
- Diet foods
- Electrical products such as light bulbs, extension cords, etc.
- Feminine hygiene products
- Financial services and products such as mortgages, credit cards, savings accounts, wire transfers, etc. (typically offered in-store by a partnering bank or other financial institution)
- Flowers
- Frozen foods and crushed ice
- Fresh produce, fruits and vegetables
- Greeting cards
- House-cleaning products
- Housewares, dishware and cooking utensils (typically limited)
- Laundry products such as detergents and fabric softeners
- Lottery tickets (where operational and legal)
- Luggage items (typically limited)
- Meats, fish and seafoods (some stores may offer live fish and seafood items from aquarium tanks)
- Medicines and first aid items (primarily over-the-counter drugs, although many supermarkets also have an on-site pharmacy) (as laws allow)
- Nonalcoholic beverages such as soft drinks, juices, and bottled water (some stores may have a juice bar that prepares ready-to-drink freshly squeezed juices, smoothies, etc.)
- Personal hygiene and grooming products
- Pet foods and products
- Seasonal items and decorations
- Snack foods
- Tea and coffee (some stores may have a commercial-style grinder, typically available for self-service, or a staffed coffee bar that prepares ready-to-drink coffee and tea beverages)
- Toys and novelties

In some countries, the range of supermarket merchandise is more strictly focused on food products, although the range of goods for sale is expanding in many locations as typical store sizes continue to increase globally.

Typical Store Architecture

While branding and store advertising will differ from company to company, the layout of a supermarket remains virtually unchanged. Although big companies spend time giving consumers a pleasant shopping experience, the design of a supermarket is directly connected to the in-store marketing that supermarkets must conduct in order to get shoppers to spend more money while there.

A Kroger store, Kroger of the Villages, in Hedwig Village, Texas (Greater Houston)

Every aspect of the store is mapped out and attention is paid to color, wording and even surface texture. The overall layout of a supermarket is a visual merchandising project that plays a major role. Stores can creatively use a layout to alter customers' perceptions of the atmosphere. Alternatively, they can enhance the store's atmospherics through visual communications (signs and graphics), lighting, colors, and even scents. For example, to give a sense of the supermarket being healthy, fresh produce is deliberately located at the front of the store. In terms of bakery items, supermarkets usually dedicate 30 to 40 feet of store space to the bread aisle.

The exterior of a supermarket in Norrköping, Sweden

Supermarkets are designed to "give each product section a sense of individual difference and this is evident in the design of what are called the anchor departments; fresh produce, dairy, delicatessen, meat and the bakery". Each section has different floor coverings, style, lighting and sometimes even individual services counters to allow shoppers to feel as if there are a number of markets within this one supermarket.

Exterior of a supermarket in Kulim, Kedah, Malaysia.

Marketers use well researched techniques to try to control purchasing behaviour. The layout of a supermarket is considered by some to consist of a few rules of thumb and three layout principles. The high-draw products are placed in separate areas of the store to keep drawing the consumer through the store. High impulse and high margin products are placed in the most predominant areas to grab attention. Power products are placed on both sides of the aisle to create increased product awareness, and end caps are used to receive high exposure of a certain product whether on special, promotion or in a campaign, or a new line.

The first principle of layout is circulation. Circulation is created by arranging product so the supermarket can control the traffic flow of the consumer. Along this path there will be high-draw, high-impulse items that will influence the consumer to purchase which he or she did not intend. Service areas such as rest rooms are placed in a location which draws the consumer past certain products to create extra buys. Necessity items such as bread and milk are found at the rear of the store to increase the start of circulation. Cashiers' desks are placed in a position to promote circulation. The entrance will be on the right-hand side because research has shown that consumers who travel in a counter-clockwise direction spend more.

The second principle of layout is coordination. Coordination is the organised arrangement of product that promotes sales. Products such as fast-selling and slow-selling lines are placed in strategic positions in aid of the overall sales plan. Managers sometimes place different items in fast-selling places to increase turnover or to promote a new line.

The third principle is consumer convenience. The layout of a supermarket is designed to create a high degree of convenience to the consumer to make the shopping experience pleasant. This is done through the character of merchandising and product placement. There are many different ideas and theories in relation to layout and how product layout can influence the purchases made.

One theory suggests that certain products are placed together or near one another that are of a similar or complementary nature to increase the average customer spend. This strategy is used by to create cross-category sales similarity. In other words, the toothpaste is next to or adjacent the tooth brushes and the tea and coffee are down the same isle as the sweet biscuits. These products complement one another and placing them near is one way marketers try to increase purchases.

Consumer psychologists suggest that most buyers tend to enter the store and shop to their right first. Some supermarkets therefore choose to place the entrance to the left-hand side as the consumer will likely turn right upon entry, and this allows the consumer to do a full anticlockwise circle around the store before returning to the checkouts. This suggests that supermarket marketers should use this theory to their advantage by placing their temporary displays of products on the right-hand side to entice you to make an unplanned purchase. Furthermore, aisle ends are extremely popular with product manufacturers, who pay top dollar to have their products located there. These aisle ends are used to lure customers into making a snap purchase and to also entice them to shop down the aisle. The most obvious place supermarket layout influences consumers is at the checkout. Small displays of chocolates, magazines and drinks are located at each checkout to tempt shoppers while they wait to be served.

Criticisms

- Supermarkets often generate much food waste. With present technologies (biomethanation units), a huge amount of food waste can be avoided, and may increase revenue for the store owners, as processed food waste can be reused again and reduce electricity bills for the supermarket.
- Supermarkets can generally sell products at lower prices than traditional corner shops and markets due to higher volume throughput. This has led to small businesses losing customers and closing in many areas, which can be seen as an adverse effect on the local infrastructure.
- In the United States, major-brand supermarkets often demand slotting fees from suppliers in exchange for premium shelf space or better positioning (such as at eye-level, on the checkout aisle or at a shelf's "end cap"). This extra supplier cost (up to $30,000 per brand for a chain for each individual SKU) may be reflected in the cost of the products offered. Some critics have questioned the ethical and legal propriety of slotting fee payments and their effect on smaller suppliers.
- Britain's supermarkets have been accused of squeezing prices to farmers, forcing small shops out of business, and often favouring imports over British produce.
- In New Zealand, supermarkets have been accused of buying fresh produce from growers at low prices and selling with high mark-ups, sometimes as high as 500%.

7

Foodservice: An Overview

Eating out at restaurants, cafes, bistros, eateries etc. can be credited to the French Revolution. But eating a meal outside the comfort of your home was in place even before in the form of taverns and inns which served its clientele food and beverages. This chapter explores topics like foodservice, foodservice distributor and casting. The major components of food industry are discussed in this chapter.

Foodservice

Foodservice (US English) or catering industry (British English) defines those businesses, institutions, and companies responsible for any meal prepared outside the home. This industry includes restaurants, school and hospital cafeterias, catering operations, and many other formats.

The companies that supply foodservice operators are called foodservice distributors. Foodservice distributors sell goods like small wares (kitchen utensils) and foods. Some companies manufacture products in both consumer and foodservice versions. The consumer version usually comes in individual-sized packages with elaborate label design for retail sale. The foodservice version is packaged in a much larger industrial size and often lacks the colorful label designs of the consumer version.

Statistics

The food system, including food service and food retailing supplied $1.24 trillion worth of food in 2010 in the US, $594 billion of which was supplied by food service facilities, defined by the USDA as any place which prepares food for immediate consumption on site, including locations that are not primarily engaged in dispensing meals such as recreational facilities and retail stores. Full-service and Fast-food restaurants account for 77% of all foodservice sales, with full service restaurants accounting for just slightly more than fast food in 2010. The shifts in the market shares between fast food and full-service restaurants to market demand changes the offerings of both foods and services of both types of restaurants.

According to the National Restaurant Association a growing trend among US consumers for the food service industry is global cuisine with 66% of US consumers eating more widely in 2015 than in 2010, 80% of consumers eating 'ethnic' cuisines at least once a month, and 29% trying a new 'ethnic' cuisine within the last year.

The Foodservice distributor market size is as of 2015 $231 billion in the US; the national broadline market is controlled by US Foods and Sysco which combined have 60-70% share of the market and were blocked from merging by the FTC for reasons of market power.

Health Concerns

Foodservice tends to be, on average, higher in calories and lower in key nutrients than foods prepared at home. Most restaurants, including fast food, have added more salads and fruit offerings and either by choice or in response to local legislation provided nutrition labeling.

In the US the FDA is moving towards establishing uniform guidelines for fast food and restaurant labeling, proposed rules were published in 2011 and final regulations published on 1 December 2014 which supersede State and local menu-labeling provisions, going into effect 1 December 2015. Research has shown that the new labels may influence consumer choices, but primarily if it provides unexpected information and that health-conscious consumers are resistant to changing behaviors based on menu labeling Fast food restaurants are expected by the ERS to do better under the new menu labeling than full service restaurants as full service restaurants tend to offer much more calorie dense foods, with 50% of fast food meals being between 400 and 800 calories and less than 20% above 1000 calories, in contrast full service restaurants 20% of meals are above 1,400 calories. When consumers are aware of the calorie counts at full service restaurants 20% choose lower calorie options and consumers also reduce their calorie intake over the rest of the day.

Eating one meal away from home each week translates to 2 extra pounds each year or a daily increase of 134 calories and a decrease in diet quality by 2 points on the Healthy Eating Index.

Table Service

Table service is food service served to the customer's table by waiters and waitresses, also known as "servers". Table service is common in most restaurants, while for some fast food restaurants counter service is the common form. For pubs and bars, counter service is the norm in the United Kingdom. With table service, the customer generally pays at the end of meal. Various methods of table service can be provided, such as silver service.

Gueridon Service

Gueridon service is a form of food service provided by restaurants to their customers. This type of service encompasses preparing food (primarily salads, main dishes such as beef tartar], or desserts) in direct view of the guests, using a "gueridon". A gueridon typically consists of a trolley that is equipped to transport, prepare, cook and serve food. There is a gas hob, chopping board, cutlery drawer, cold store (depending on the trolley type) and general working area.

Foodservice Distributor

A food service distributor is a company that provides food and non-food products to restaurants, cafeterias, industrial caterers, hospitals and nursing homes.

A food service distributor functions as an intermediary between food manufacturers and the food service operator (usually a chef, food service director, food and beverage manager, and independent food preparation businesses operator owners.) The distributor purchases, stores, sells, and delivers those products, providing food service operators with access to items from a wide variety of manufacturers. Food service distributors procure pallets and bulk inventory quantities that are broken down to case and sometimes unit quantities for the food service operator. Most food service operators purchase from a range of local, specialty, and broadline food service distributors on a daily or weekly basis.

Often a food manufacturer may hire a food brokerage company to represent the manufacturer in a local market. The broker helps the food manufacturer market its products through the food service distribution system, which ranges from getting items stocked at the distributor to working with operators to purchase items from the distributor. At the same time, distributor sales teams work to market products directly to operator customers.

A broadline distributor services a wide variety of accounts with a wide variety of products, while a system distributor stocks a narrow array of products for specific customers, such as restaurant chains. A broadline distributor may carry up to 15,000 different items for purchase and operate sophisticated warehouse and transportation operations.

A small food service distributor is often referred to as a "wagon-jobber". These wagon-jobbers will purchase food in bulk and deliver small quantities to independent retail stores keeping their shelves stocked. Independent distributors and jobbers service independent convenience store markets, bodegas and niche grocery stores. While these distributors are unorganized, networks of independent distributors and wagon jobbers have emerged to give these jobbers the ability to identify trends in the market.

It is estimated by food industry research firm Technomics that approximately 225 million meals are eaten away from home each day in the U.S. This includes both restaurant and non-commercial eating places. The International Foodservice Distributors Association estimates that food service distributors in the U.S., as a daily average, deliver approximately 27 million cases of food and other products.

Food service distribution companies can range in size from a one-truck operation to larger corporations. There are many independent broadline food service distribution companies that service chain and multi-unit restaurants based on master distribution agreements with national food service groups. These groups provide distributor members procurement capabilities rival the purchasing power of largest distributors. These distributor groups also provide distributor members group private label brands as well as marketing and quality assurance services.

In the US, the industry is highly fragmented, with Sysco capturing 17% of the market, US Foods with about 9%, PFG with 5%, and Gold Star Foods playing a large part as well. The rest are spread across a host of smaller, regional players.

Redistribution

In the food *redistribution* model, a redistributor will purchase in truckload quantities from many food manufacturers and warehouse these products for its customers. Individual distributors (typically smaller in size and service area) can then purchase items across multiple manufacturers' on one easy to place order from the redistributor.

The redistribution model affords smaller distributors who are unable to purchase direct truckloads an opportunity to purchase from a non-competitor in less than truckload (LTL) quantities, giving them the ability to compete against larger distributors in their territory. Typically it is the smaller distributor that services the independent non-chain retail outlets often overlooked by the larger distributors.

Casting

Casting is a manufacturing process in which a liquid material is usually poured into a mold, which contains a hollow cavity of the desired shape, and then allowed to solidify. The solidified part is also known as a *casting*, which is ejected or broken out of the mold to complete the process. Casting materials are usually metals or various *cold setting* materials that cure after mixing two or more components together; examples are epoxy, concrete, plaster and clay. Casting is most often used for making complex shapes that would be otherwise difficult or uneconomical to make by other methods.

Judenplatz Holocaust Memorial (Nameless Library), by Rachel Whiteread. Concrete cast of books on library shelves turned inside out.

Casting iron in a sand mold

Molten metal before casting

Casting is a 6000-year-old process. The oldest surviving casting is a copper frog from 3200 BC.

Types

Metal

In metalworking, metal is heated until it becomes liquid and is then poured into a mold. The mold is a hollow cavity that includes the desired shape, but the mold also includes runners and risers that enable the metal to fill the mold. The mold and the metal are then cooled until the metal solidifies. The solidified part (the casting) is then recovered from the mold. Subsequent operations remove excess material caused by the casting process (such as the runners and risers).

Plaster, Concrete, or Plastic Resin

Plaster and other chemical curing materials such as concrete and plastic resin may be cast using single-use *waste* molds as noted above, multiple-use 'piece' molds, or molds made of small rigid pieces or of flexible material such as latex rubber (which is in turn supported by an exterior mold). When casting plaster or concrete, the material surface is flat and lacks transparency. Often topical treatments are applied to the surface. For example, painting and etching can be used in a way that give the appearance of metal or stone. Alternatively, the material is altered in its initial casting process and may contain colored sand so as to give an appearance of stone. By casting concrete, rather than plaster, it is possible to create sculptures, fountains, or seating for outdoor use. A simulation of high-quality marble may be made using certain chemically-set plastic resins (for example epoxy or polyester) with powdered stone added for coloration, often with multiple colors worked in. The latter is a common means of making washstands, washstand tops and shower stalls, with the skilled working of multiple colors resulting in simulated staining patterns as is often found in natural marble or travertine.

Fettling

Raw castings often contain irregularities caused by seams and imperfections in the molds, as well as access ports for pouring material into the molds. The process of cutting, grinding, shaving or sanding away these unwanted bits is called "fettling". In modern times robotic processes have been developed to perform some of the more repetitive parts of the fettling process, but historically fettlers carried out this arduous work manually, and often in conditions dangerous to their health.

Fettling can add significantly to the cost of the resulting product, and designers of molds seek to minimize it through the shape of the mold, the material being cast, and sometimes by including decorative elements.

Casting Process Simulation

Casting process simulation uses numerical methods to calculate cast component quality considering mold filling, solidification and cooling, and provides a quantitative prediction of casting mechanical properties, thermal stresses and distortion. Simulation accurately describes a cast component's quality up-front before production starts. The casting rigging can be designed with respect to the required component properties. This has benefits beyond a reduction in pre-production sampling, as the precise layout of the complete casting system also leads to energy, material, and tooling savings.

The software supports the user in component design, the determination of melting practice and casting methoding through to pattern and mold making, heat treatment, and finishing. This saves costs along the entire casting manufacturing route.

Casting process simulation was initially developed at universities starting from the early '70s, mainly in Europe and in the U.S., and is regarded as the most important innovation in casting technology over the last 50 years. Since the late '80s, commercial programs (such as AutoCAST and MAGMA) are available which make it possible for foundries to gain new insight into what is happening inside the mold or die during the casting process.

References

- Degarmo, E. Paul; Black, J T.; Kohser, Ronald A. (2003), Materials and Processes in Manufacturing (9th ed.), Wiley, p. 277, ISBN 0-471-65653-4
- Gordon Elliott (2006). Aspects of Ceramic History: A Series of Papers Focusing on the Ceramic Artifact As Evidence of Cultural and Technical Development. Gordon Elliott. pp. 52–. ISBN 978-0-9557690-0-9.

- B. Ravi (1 January 2005). Metal Casting: Computer-Aided Design and Analysis. PHI Learning Pvt. Ltd. pp. 92–. ISBN 978-81-203-2726-9.
- James T. Frane (1994). Craftsman's Illustrated Dictionary of Construction Terms. Craftsman Book Company. pp. 126–. ISBN 978-1-57218-008-6.
- Jane L. Bassett; Peggy Fogelman; David A. Scott; Ronald C. Schmidtling (2008). The Craftsman Revealed: Adriaen de Vries, Sculptor in Bronze. Getty Publications. pp. 269–. ISBN 978-0-89236-919-5.
- Stewart, Hayden; Morrison, Rosanna Mentzer. "New Regulations Will Inform Consumers About Calories in Restaurant Foods". USDA ERS. Retrieved 27 August 2015.
- Todd, Jessica E.; Mancino, Lisa; Lin, Biing-Hwan. "The Impact of Food Away From Home on Adult Diet Quality - Report Summary". USDA ERS. Retrieved 27 August 2015.

8

Food Technology: A Comprehensive Study

All the food production methods and technologies can be listed under food technology. It comes as no surprise that with the leaps and bounds that science has made in the last few decades that food technology is still evolving and incorporating newer elements. This chapter discusses developments in food technology like powdered milk, freeze-drying, decaffeination and ultra-high temperature processing. A section of the content focuses on food engineering and food packaging techniques and machines.

Food Technology

Food Technology is a branch of food science that deals with the production processes that make foods.

Bakery at the Faculty of Food Technology, Latvia University of Agriculture

Early scientific research into food technology concentrated on food preservation. Nicolas Appert's development in 1810 of the canning process was a decisive event. The process wasn't called canning then and Appert did not really know the principle on which his process worked, but canning has had a major impact on food preservation techniques.

Louis Pasteur's research on the spoilage of wine and his description of how to avoid spoilage in 1864 was an early attempt to apply scientific knowledge to food handling. Besides research into wine spoilage, Pasteur researched the production of alcohol, vinegar, wines and beer, and the souring of milk. He developed pasteurization—the process of heating milk and milk products to destroy food spoilage and disease-producing organisms. In his research into food technology, Pasteur became the pioneer into bacteriology and of modern preventive medicine.

The food technology room at Marling School in Stroud, Gloucestershire

Developments

Developments in food technology have contributed greatly to the food supply and have changed our world. Some of these developments are:

Freeze-dried coffee, a form of instant coffee

- Instantized Milk Powder - D.D. Peebles (U.S. patent 2,835,586) developed the first instant milk powder, which has become the basis for a variety of new products that are rehydratable. This process increases the surface area of the powdered product by partially rehydrating spray-dried milk powder.
- Freeze-drying - The first application of freeze drying was most likely in the pharmaceutical industry; however, a successful large-scale industrial application of the process was the development of continuous freeze drying of coffee.
- High-Temperature Short Time Processing - These processes for the most part are characterized by rapid heating and cooling, holding for a short time at a relatively high temperature and filling aseptically into sterile containers.
- Decaffeination of Coffee and Tea - Decaffeinated coffee and tea was first developed on a commercial basis in Europe around 1900. The process is described in U.S. patent 897,763. Green coffee beans are treated with water, heat and solvents to remove the caffeine from the beans.

- Process optimization - Food Technology now allows production of foods to be more efficient, Oil saving technologies are now available on different forms. Production methods and methodology have also become increasingly sophisticated.

Consumer Acceptance

In the past, consumer attitude towards food technologies was not common talk and was not important in food development. Nowadays the food chain is long and complicated, foods and food technologies are diverse; consequently the consumers are uncertain about the food quality and safety and find it difficult to orient themselves to the subject. That is why consumer acceptance of food technologies is an important question. However, in these days acceptance of food products very often depends on potential benefits and risks associated with the food. This also includes the technology the food is processed with. Attributes like “uncertain”, “unknown” or “unfamiliar” are associated with consumers’ risk perception and consumer very likely will reject products linked to these attributes. Especially innovative food processing technologies are connected to these characteristics and are perceived as risky by consumers

Acceptance of the different food technologies is very different. Whereas pasteurisation is well recognised, high pressure treatment or even microwaves are perceived as risky very often. In studies done within Hightech Europe project, it was found that traditional technologies were well accepted in contrast to innovative technologies.

Consumers form their attitude towards innovative food technologies by three main factors mechanisms. First, knowledge or beliefs about risks and benefits which are correlated with the technology. Second, attitudes are based on their own experience and third, based on higher order values and beliefs.

Acceptance of innovative technologies can be improved by providing non-emotional and concise information about these new technological processes methods. According to a study made by HighTech project also written information seems to have higher impact than audio-visual information on the consumer in case of sensory acceptance of products processed with innovative food technologies.

Developments in Food Technology

Powdered Milk

Powdered milk or dried milk is a manufactured dairy product made by evaporating milk to dryness. One purpose of drying milk is to preserve it; milk powder has a far longer shelf life than liquid milk and does not need to be refrigerated, due to its low moisture content. Another purpose is to reduce its bulk for economy of transportation. Powdered milk and dairy products include such items as dry whole milk, nonfat (skimmed) dry milk, dry buttermilk, dry whey products and dry dairy blends. Many dairy products exported conform to standards laid out in Codex Alimentarius. Many forms of milk powder are traded on exchanges.

Powdered milk is used for food and health (nutrition), and also in biotechnology (saturating agent).

Powdered milk

History and Manufacture

While Marco Polo wrote of Mongolian Tatar troops in the time of Kublai Khan who carried sun-dried skimmed milk as "a kind of paste", the first modern production process for dried milk was invented by the Russian physician Osip Krichevsky in 1802. The first commercial production of dried milk was organized by the Russian chemist M. Dirchoff in 1832. In 1855, T.S. Grimwade took a patent on a dried milk procedure, though a William Newton had patented a vacuum drying process as early as 1837.

Modified dry whole milk, fortified with vitamin D. This is the original container from 1947, provided by the Ministry of Food in London, England

In modern times, powdered milk is usually made by spray drying nonfat skimmed milk, whole milk, buttermilk or whey. Pasteurized milk is first concentrated in an evaporator to approximately 50% milk solids. The resulting concentrated milk is then sprayed into a heated chamber where the water almost instantly evaporates, leaving fine particles of powdered milk solids.

Alternatively, the milk can be dried by drum drying. Milk is applied as a thin film to the surface of a heated drum, and the dried milk solids are then scraped off. However, powdered milk made this way tends to have a cooked flavor, due to caramelization caused by greater heat exposure.

Another process is freeze drying, which preserves many nutrients in milk, compared to drum drying.

The drying method and the heat treatment of the milk as it is processed alters the properties of the milk powder, such as its solubility in cold water, its flavor, and its bulk density.

Food and Health Uses

Powdered milk is frequently used in the manufacture of infant formula, confectionery such as chocolate and caramel candy, and in recipes for baked goods where adding liquid milk would render the product too thin. Powdered milk is also widely used in various sweets such as the famous Indian milk balls known as gulab jamun and popular Indian sweet delicacy (sprinkled with desiccated coconut) known as Chum chum (made with skim milk powder).

Incolac powdered milk

Powdered milk is also a common item in UN food aid supplies, fallout shelters, warehouses, and wherever fresh milk is not a viable option. It is widely used in many developing countries because of reduced transport and storage costs (reduced bulk and weight, no refrigerated vehicles). Like other dry foods, it is considered nonperishable, and is favored by survivalists, hikers, and others requiring nonperishable, easy-to-prepare food.

Because of its resemblance to cocaine and other drugs, powdered milk is sometimes used in filmmaking as a non-toxic prop that may be insufflated.

Reconstitution

The weight of nonfat dry milk (NFDM) to use is about 10% of the water weight. Alternatively, one cup of potable fluid milk from powdered milk requires one cup of potable water and one-third cup of powdered milk.

Nutritional Value

Milk powders contain all twenty-one standard amino acids, the building blocks of proteins, and are high in soluble vitamins and minerals. According to USAID, the typical average amounts of major nutrients in the unreconstituted nonfat dry milk are (by weight) 36% protein, 52% carbohydrates (predominantly lactose), calcium 1.3%, potassium 1.8%. Whole milk powder, on the other hand, contains on average 25-27% protein, 36-38% carbohydrates, 26-40% fat, and 5-7% ash (minerals). However, inappropriate storage conditions such as high relative humidity and high ambient temperature can significantly degrade the nutritive value of milk powder.

Commercial milk powders are reported to contain oxysterols (oxidized cholesterol) in higher amounts than in fresh milk (up to 30 μg/g, versus trace amounts in fresh milk). Oxysterols are derivatives of cholesterol that are produced either by free radicals or by enzymes. Some free radicals-derived oxysterols have been suspected of being initiators of atherosclerotic plaques. For comparison, powdered eggs contain even more oxysterols, up to 200 μg/g.

Export Market

European production of milk powder is estimated around 800,000 tons of which the main volume is exported in bulk packing or consumer packs.

National household dried machine skimmed milk. This was U.S.-produced dry milk for food export in June 1944.

Brands on the market include "Nido", from the company Nestlé, "Incolac" from the company Belgomilk, and "Dutch Lady" from FrieslandCampina.

Adulteration

In the 2008 Chinese milk scandal, melamine adulterant was found in Sanlu infant formula, added to fool tests into reporting higher protein content. Thousands became ill, and some children died, after consuming the product.

In August 2013, China temporarily suspended all milk powder imports from New Zealand, after a scare where botulism-causing bacteria was falsely detected in several batches of New Zealand-produced whey protein concentrate. As a result of the product recall, the New Zealand dollar slipped significantly based on expected losses in sales from this single commodity.

Use in Biotechnology

Fat-free powdered milk is used as a saturating agent to block nonspecific binding sites on supports like blotting membranes (Nitrocellulose, PVDF, Nylon), preventing binding of further detection reagents and subsequent background. It may be referred as Blotto. The major protein of milk, casein, is responsible for most of the binding site saturation effect.

Freeze-drying

Freeze-drying—technically known as lyophilisation, lyophilization, or cryodesiccation—is a dehydration process typically used to preserve a perishable material or make the material more convenient for transport. Freeze-drying works by freezing the material and then reducing the surrounding pressure to allow the frozen water in the material to sublimate directly from the solid phase to the gas phase.

Freeze dried ice cream

A benchtop manifold freeze-drier

History

The process of freeze-drying was invented in 1906 by Arsène d'Arsonval and his assistant Frèdèric Bordas at the laboratory of biophysics of Collège de France in Paris. In 1911 Downey Harris and Shackle developed the freeze-drying method of preserving live rabies virus which eventually led to development of the first antirabies vaccine.

Modern freeze-drying was developed during WWII. Blood serum being sent to Europe from the US for medical treatment of the wounded required refrigeration, but because of the lack of simultaneous refrigeration and transport, many serum supplies were spoiling before reaching their intended recipients. The freeze-drying process was developed as a commercial technique that enabled serum to be rendered chemically stable and viable without having to be refrigerated. Shortly thereafter, the freeze-dry process was applied to penicillin and bone, and lyophilization became recognized as an important technique for preservation of biologicals. Since that time, freeze-drying has been used as a preservation or processing technique for a wide variety of products. These applications include the following but are not limited to: the processing of food, pharmaceuticals, and diagnostic kits; the restoration of water damaged documents; the preparation of river-bottom sludge for hydrocarbon analysis; the manufacturing of ceramics used in the semiconductor industry; the production of synthetic skin; the manufacture of sulfur-coated vials; and the restoration of historic/reclaimed boat hulls.

Stages

There are four stages in the complete drying process: pretreatment, freezing, primary drying, and secondary drying.

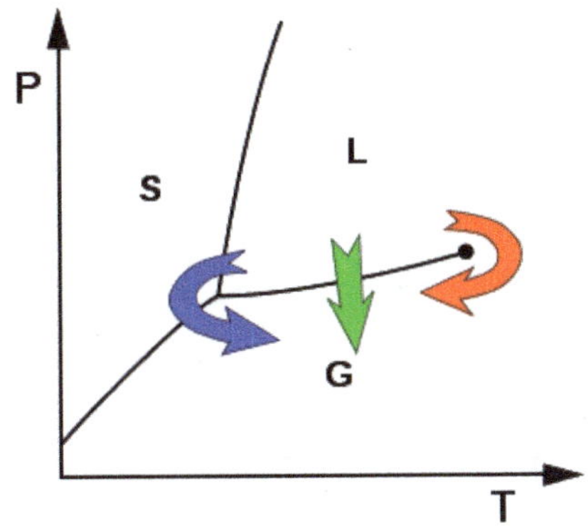

In a typical phase diagram, the boundary between gas and liquid runs from the triple point to the critical point. Freeze-drying (blue arrow) brings the system around the triple point, avoiding the direct liquid-gas transition seen in ordinary drying time (green arrow).

Pretreatment

Pretreatment includes any method of treating the product prior to freezing. This may include concentrating the product, formulation revision (i.e., addition of components to increase stability, preserve appearance, and/or improve processing), decreasing a high-vapor-pressure solvent, or increasing the surface area. In many instances the decision to pretreat a product is based on theoretical knowledge of freeze-drying and its requirements, or is demanded by cycle time or product quality considerations.

Freezing

In a lab, this is often done by placing the material in a freeze-drying flask and rotating the flask in a bath, called a shell freezer, which is cooled by mechanical refrigeration, dry ice and methanol, or liquid nitrogen. On a larger scale, freezing is usually done using a freeze-drying machine. In this step, it is important to cool the material below its triple point, the lowest temperature at which the solid and liquid phases of the material can coexist. This ensures that sublimation rather than melting will occur in the following steps. Larger crystals are easier to freeze-dry. To produce larger crystals, the product should be frozen slowly or can be cycled up and down in temperature. This cycling process is called annealing. However, in the case of food, or objects with formerly-living cells, large ice crystals will break the cell walls (a problem discovered, and solved, by Clarence Birdseye), resulting in the destruction of more cells, which can result in increasingly poor texture and nutritive content. In this case, the freezing is done rapidly, in order to lower the material to below its eutectic point quickly, thus avoiding the formation of ice crystals. Usually, the freezing temperatures are between −50 °C and −80 °C (-58 °F and -112 °F) . The freezing phase is the most critical in the whole freeze-drying process, because the product can be spoiled if improperly done.

Amorphous materials do not have a eutectic point, but they do have a critical point, below which the product must be maintained to prevent melt-back or collapse during primary and secondary drying.

Primary Drying

During the primary drying phase, the pressure is lowered (to the range of a few millibars), and enough heat is supplied to the material for the ice to sublime. The amount of heat necessary can be calculated using the sublimating molecules' latent heat of sublimation. In this initial drying phase, about 95% of the water in the material is sublimated. This phase may be slow (can be several days in the industry), because, if too much heat is added, the material's structure could be altered.

In this phase, pressure is controlled through the application of partial vacuum. The vacuum speeds up the sublimation, making it useful as a deliberate drying process. Furthermore, a cold condenser chamber and/or condenser plates provide a surface(s) for the water vapour to re-solidify on. This condenser plays no role in keeping the material frozen; rather, it prevents water vapor from reaching the vacuum pump, which could degrade the pump's performance. Condenser temperatures are typically below −50 °C (−58 °F).

It is important to note that, in this range of pressure, the heat is brought mainly by conduction or radiation; the convection effect is negligible, due to the low air density.

Secondary Drying

The secondary drying phase aims to remove unfrozen water molecules, since the ice was removed in the primary drying phase. This part of the freeze-drying process is governed by the material's adsorption isotherms. In this phase, the temperature is raised higher than in the primary drying phase, and can even be above 0 °C, to break any physico-chemical interactions that have formed between the water molecules and the frozen material. Usually the pressure is also lowered in this stage to encourage desorption (typically in the range of microbars, or fractions of a pascal). However, there are products that benefit from increased pressure as well.

After the freeze-drying process is complete, the vacuum is usually broken with an inert gas, such as nitrogen, before the material is sealed.

At the end of the operation, the final residual water content in the product is extremely low, around 1% to 4%.

Properties of Freeze-dried Products

If a freeze-dried substance is sealed to prevent the reabsorption of moisture, the substance may be stored at room temperature without refrigeration, and be protected against spoilage for many years. Preservation is possible because the greatly reduced water content inhibits the action of microorganisms and enzymes that would normally spoil or degrade the substance.

Freeze-drying also causes less damage to the substance than other dehydration methods using higher temperatures. Freeze-drying does not usually cause shrinkage or toughening of the material being dried. In addition, flavours, smells and nutritional content generally remain unchanged, making the process popular for preserving food. However, water is not the only chemical capable of sublimation, and the loss of other volatile compounds such as acetic acid (vinegar) and alcohols can yield undesirable results.

Freeze-dried products can be rehydrated (reconstituted) much more quickly and easily because the process leaves microscopic pores. The pores are created by the ice crystals that sublimate, leaving gaps or pores in their place. This is especially important when it comes to pharmaceutical uses. Freeze-drying can also be used to increase the shelf life of some pharmaceuticals for many years.

Protectants

Similar to cryoprotectants, some molecules protect freeze-dried material. Known as lyoprotectants, these molecules are typically polyhydroxy compounds such as sugars (mono-, di-, and polysaccharides), polyalcohols, and their derivatives. Trehalose and sucrose are natural lyoprotectants. Trehalose is produced by a variety of plant (for example selaginella and arabidopsis thaliana), fungi, and invertebrate animals that remain in a state of suspended animation during periods of drought (also known as anhydrobiosis).

Applications

Pharmaceutical and Biotechnology

Lyophilized 5% w/v sucrose cake in a pharmaceutical glass vial

Pharmaceutical companies often use freeze-drying to increase the shelf life of the products, such as live virus vaccines, biologics and other injectables. By removing the water from the material and sealing the material in a glass vial, the material can be easily stored, shipped, and later reconstituted to its original form for injection. Another example from the pharmaceutical industry is the use of freeze drying to produce tablets or wafers, the advantage of which is less excipient as well as a rapidly absorbed and easily administered dosage form.

Freeze-dried pharmaceutical products are produced as lyophilized powders for reconstitution in vials and more recently in prefilled syringes for self-administration by a patient. Many biopharmaceutical products based on therapeutic proteins such as monoclonal antibodies require lyophilization for stability. Examples of lyophilized biopharmaceuticals include blockbuster drugs such as Etanercept (Enbrel by Pfizer), Infliximab (Remicade by Janssen Biotech), Rituximab and Trastuzumab (Herceptin by Genentech).

Freeze-drying is also used in manufacturing of raw materials for pharmaceutical products. Active Pharmaceutical Product Ingredients (APIs) are lyophilized to achieve chemical stability under room temperature storage. Bulk lyophilization of APIs is typically conducted using trays instead of glass vials.

Dry powders of probiotics are often produced by bulk freeze-drying of live microorganisms such as Lactic acid bacteria and Bifidobacteria.

Food and Agriculture-based Industries

Although freeze-drying is used to preserve food, its earliest use in agriculturally based industries was in processing of crops such as peanuts/groundnuts and tobacco in the early 1970s. Because heat, commonly used in crop and food processing, invariably alters the structure and chemistry of the product, the main objective of freeze-drying is to avoid heat and thus preserve the structural and chemical integrity/composition with little or no alteration. Therefore, freeze-dried crops and foods are closest to the natural composition with respect to structure and chemistry. The process came to wide public attention when it was used to create freeze-dried ice cream, an example of astronaut food. It is also widely used to produce essences or flavourings to add to food.

Freeze dried bacon bars

Because of its light weight per volume of reconstituted food, freeze-dried products are popular and convenient for hikers. More dried food can be carried per the same weight of wet food, and remains in good condition for longer than wet food, which tends to spoil quickly. Hikers reconstitute the food with water available at point of use.

Instant coffee is sometimes freeze-dried, despite the high costs of the freeze-driers used. The coffee is often dried by vaporization in a hot air flow, or by projection onto hot metallic plates. Freeze-dried fruits are used in some breakfast cereal or sold as a snack, and are an especially popular snack choice among toddlers, preschoolers, hikers and dieters, as well as being used by some pet owners as a treat for pet birds. Most commercial freezing is done either in cold air kept in motion by fans (blast freezing) or by placing the foodstuffs in packages or metal trays on refrigerated surfaces (contact freezing).

Culinary herbs, vegetables (such as vitamin-rich spinach and watercress), the temperature sensitive baker`s yeast suspension and the nutrient-rich pre-boiled rice can also be freeze-dried. During three hours of drying the spinach and watercress has lost over 98% of its water content, followed by the yeast suspension with 96% and the pre-boiled rice by 75%. The air-dried herbs are far more common and less expensive. Freeze dried tofu is a popular foodstuff in Japan ("Koya-dofu" or "shimi-dofu" in Japanese).

Technological Industry

In chemical synthesis, products are often freeze-dried to make them more stable, or easier to dissolve in water for subsequent use.

In bioseparations, freeze-drying can be used also as a late-stage purification procedure, because it can effectively remove solvents. Furthermore, it is capable of concentrating substances with low molecular weights that are too small to be removed by a filtration membrane. Freeze-drying is a relatively expensive process. The equipment is about three times as expensive as the equipment used for other separation processes, and the high energy demands lead to high energy costs. Furthermore, freeze-drying also has a long process time, because the addition of too much heat to the material can cause melting or structural deformations. Therefore, freeze-drying is often reserved for materials that are heat-sensitive, such as proteins, enzymes, microorganisms, and blood plasma. The low operating temperature of the process leads to minimal damage of these heat-sensitive products.

In nanotechnology, freeze-drying is used for nanotube purification to avoid aggregation due to capillary forces during regular thermal vaporization drying.

Other Uses

Organizations such as the Document Conservation Laboratory at the United States National Archives and Records Administration (NARA) have done studies on freeze-drying as a recovery method of water-damaged books and documents. While recovery is possible, restoration quality depends on the material of the documents. If a document is made of a variety of materials, which have different absorption properties, expansion will occur at a non-uniform rate, which could lead to deformations. Water can also cause mold to grow or make inks bleed. In these cases, freeze-drying may not be an effective restoration method.

In bacteriology freeze-drying is used to conserve special strains.

In high-altitude environments, the low temperatures and pressures can sometimes produce natural mummies by a process of freeze-drying.

Advanced ceramics processes sometimes use freeze-drying to create a formable powder from a sprayed slurry mist. Freeze-drying creates softer particles with a more homogeneous chemical composition than traditional hot spray drying, but it is also more expensive.

Freeze-drying is also used for floral preservation. Wedding bouquet preservation has become very popular with brides who want to preserve their wedding day flowers

A new form of burial which previously freeze-dries the body with liquid nitrogen has been developed by the Swedish company Promessa Organic AB, which puts it forward as an environmentally friendly alternative to traditional casket and cremation burials.

Equipment

There are essentially three categories of freeze-dryers: the manifold freeze-dryer, the rotary freeze-dryer and the tray style freeze-dryer. Two components are common to all types of freeze-dryers: a vacuum pump to reduce the ambient gas pressure in a vessel containing the substance to be dried and a condenser to remove the moisture by condensation on a surface cooled to −40 to −80 °C (−40 to −112 °F). The manifold, rotary and tray type freeze-dryers differ in the method by which the dried substance is interfaced with a condenser. In manifold freeze-dryers a short usually circular tube is used to connect multiple containers with the dried product to a condenser. The rotary and tray freeze-dryers have a single large reservoir for the dried substance.

Unloading trays of freeze-dried material from a small cabinet-type freeze-dryer

Rotary freeze-dryers are usually used for drying pellets, cubes and other pourable substances. The rotary dryers have a cylindrical reservoir that is rotated during drying to achieve a more uniform drying throughout the substance. Tray style freeze-dryers usually have rectangular reservoir with shelves on which products, such as pharmaceutical solutions and tissue extracts, can be placed in trays, vials and other containers.

Manifold freeze-dryers are usually used in a laboratory setting when drying liquid substances in small containers and when the product will be used in a short period of time. A manifold dryer will dry the product to less than 5% moisture content. Without heat, only primary drying (removal of the unbound water) can be achieved. A heater must be added for secondary drying, which will remove the bound water and will produce a lower moisture content.

Tray style freeze-dryers are typically larger than the manifold dryers and are more sophisticated. Tray style freeze-dryers are used to dry a variety of materials. A tray freeze-dryer is used to produce the driest product for long-term storage. A tray freeze-dryer allows the product to be frozen in place and performs both primary (unbound water removal) and secondary (bound water removal) freeze-drying, thus producing the driest possible end-product. Tray freeze-dryers can dry products in bulk or in vials or other containers. When drying in vials, the freeze-dryer is supplied with a stoppering mechanism that allows a stopper to be pressed into place, sealing the vial before it is exposed to the atmosphere. This is used for long-term storage, such as vaccines.

Improved freeze-drying techniques are being developed to extend the range of products that can be freeze-dried, to improve the quality of the product, and to produce the product faster with less labor.

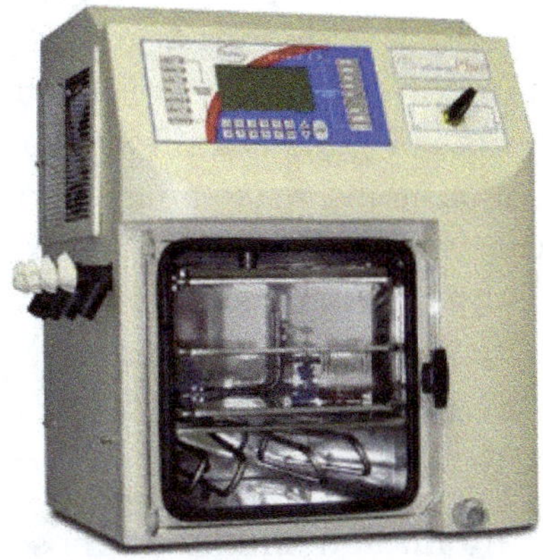

Product viewable single shelf freeze-dryer

Production freeze-drier

Decaffeination

Decaffeination is the removal of caffeine from coffee beans, cocoa, tea leaves and other caffeine-containing materials. While soft drinks which do not use caffeine as an ingredient are sometimes described as "decaffeinated", they are better termed "non-caffeinated" because decaffeinated implies that there was caffeine present at one point in time. Decaffeinated drinks contain typically 1–2% of the original caffeine content, and sometimes as much as 20%. Decaffeinated products are commonly termed decaf.

History

Coffee

Friedlieb Ferdinand Runge performed the first isolation of pure caffeine from coffee beans in 1820. He did this after the Poet Goethe requested he perform an analysis on coffee beans after seeing his work on belladonna extract. Though Runge was able to isolate the compound, he didn't learn much as to the chemistry of caffeine itself, nor did he seek to use the process commercially to produce decaffeinated coffee.

The first commercially successful decaffeination process was invented by German merchant Ludwig Roselius and co-workers in 1903 and patented in 1906. In 1903, Ludwig accidentally stumbled upon this method when his freight of coffee beans was soaked in sea water and lost much of its caffeine without losing much taste. This original decaffeination process involved steaming coffee beans with various acids or bases, then using benzene as a solvent to remove the caffeine. Coffee decaffeinated this way was sold as Kaffee HAG after the company name *Kaffee Handels-Aktien-Gesellschaft* (Coffee Trading Company) in most of Europe, as *Café Sanka* in France and later as Sanka brand coffee in the U.S.. Café HAG and Sanka are now worldwide brands of Kraft Foods. Due to health concerns regarding benzene (which is recognised today as a proven carcinogen), benzene is no longer used as a solvent commercially.

Since its inception, methods of decaffeination similar to those first developed by Roselius have continued to dominate. While Roselius used Benzene, many different solvents have since been tried after learning of the potential harmful effects of Benzene. The most prevalent solvents used to date are Dichloromethane and Ethyl Acetate, which are still used largely today.

Another variation of Roselius's method is the indirect organic solvent method. This is very similar to the process described above, only instead of treating the beans directly, water resulting from the soaking of beans is treated with solvents and the process goes on until equilibrium is reached without caffeine in the beans. This method was first mentioned in 1941, and people have made great efforts to make the process more "natural' and a true water-based process by finding ways to process the caffeine out of the water in ways that circumvents the use of organic solvents.

Another process, known as the Swiss Water Method, uses solely water and osmosis to decaffeinate beans. The use of water as the solvent to decaffeinate coffee was originally pioneered in Switzerland in 1933 and developed as a commercially viable method of decaffeination by Coffex S.A. in 1980. In 1988, the Swiss Water Method was introduced by The Swiss Water Decaffeinated Coffee Company of Burnaby, British Columbia, Canada. Norwegian chemical engineer Torunn Atteraas Garin also worked a process to remove caffeine from coffee.

Most recently, food scientists have turned to supercritical carbon dioxide as a means of decaffeination. Developed by Kurt Zosel, a scientist of the Max Plank Institute, it uses liquid CO_2 to extract caffeine and could be useful going forward because it circumvents the use of other solvents and their possible effects entirely.

Decaffeination Processes for Coffee

In the case of coffee, various methods can be used. The process is performed on unroasted (green) beans and starts with steaming of the beans. They are then rinsed with a solvent that extracts the caffeine while leaving other constituents largely unaffected. The process is repeated from 8 to 12 times until the caffeine content meets the required standard (97% of caffeine removed according to the US standard, or 99.9% caffeine-free by mass per the EU standard).

Common Characteristics of Decaffeination

In all decaffeination processes, coffee is always decaffeinated in its green, unroasted state. The greatest challenge to the decaffeination process is to try to separate only the caffeine from the coffee beans while leaving the other chemicals such as sucrose, cellulose, proteins, citric acid, tartaric acid, and formic acid at their original concentrations. This is not an easy task considering coffee contains somewhere around 1,000 chemicals that contribute to the taste and aroma. Since caffeine is a polar, water-soluble substance, water is used in all forms of decaffeination. However, water alone is not the best solution for decaffeination because it is not a selective solvent and therefore removes other soluble substances, including sugars and proteins, as well as caffeine. Therefore, most decaffeination processes use a decaffeinating agent such as methylene chloride, activated charcoal, CO_2, or ethyl acetate. These agents help speed up the process and minimize the "washed-out" effects that water alone might have on the taste of decaffeinated coffee.

Swiss Water Process

This method is different in that it does not directly or indirectly add chemicals to extract the caffeine. Rather, it relies entirely on two concepts – solubility and osmosis – to decaffeinate coffee beans. The process begins by immersing a batch of green coffee beans in very hot water in order to dissolve and extract the caffeine. The water is then drawn off and passed through an activated charcoal filter. The porosity of this filter is sized to only capture larger caffeine molecules, while allowing smaller oil and flavor molecules to pass through it. However, this extraction process will also extract desirable oils and other solids from the beans, resulting in beans with no caffeine and no flavor in one tank, and caffeine-free but with flavor water in another tank. The Swiss Water process method attempts to overcome this difficulty by first discarding the flavorless caffeine-free beans, and then reusing the flavor-rich water to remove the caffeine from a fresh batch of coffee beans. Water saturated in this way is referred to as green coffee extract or GCE. It is created using a separate batch of green coffee beans, which are immersed in water and then discarded. In a pre-loading tank, water, cane sugar and formic acid are mixed and heated and used to preload carbon filter columns. The GCE is then filtered over the columns to extract caffeine from it. A fresh batch of green coffee beans is then immersed in the GCE to remove caffeine but retain other components. The other components can be retained because the water is already saturated with flavor ingredients, therefore, the flavors in this fresh batch cannot dissolve - only caffeine moves from the coffee beans to the water. This results in decaffeination without a massive loss of flavor. The process of filtering the GCE to remove caffeine and immersing the beans is repeated until the beans are 99.9% caffeine free by mass, meeting the required standard. This process takes 8 to 10 hours.

Organic Solvent Processes

Solvents Used in Decaffeination

Given numerous health scares connected to early efforts in decaffeination using solvents such as benzene, trichloroethylene, and chloroform, the solvents of choice have become dichloromethane and ethyl acetate. Dichloromethane is able to extract the caffeine selectively and has a low boiling point. Although it is mildly toxic and carcinogenic, its use as a decaffeination agent is allowed by the US Food and Drug Administration if the residual solvent is less than 10 parts per million (ppm). Actual coffee industry practice results in residues closer to one part per million. Starting in the 1980s, ethyl acetate was introduced as a replacement to dichloromethane. Although ethyl acetate is mildly toxic, coffee that is decaffeinated with this solvent is sometimes marketed as "naturally decaffeinated" because this solvent may be obtained from a biological process such as the fermentation of sugar cane.

Direct Method

In the direct method, the coffee beans are first steamed for 30 minutes to open their pores and then repeatedly rinsed with either dichloromethane or ethyl acetate for about 10 hours to remove the caffeine. The caffeine-laden solvent is then drained away and the beans steamed to remove residual solvent.

Indirect Method

In the indirect method, beans are first soaked in hot water for several hours, in essence making a strong pot of coffee. Then the beans are removed and either dichloromethane or ethyl acetate is used to extract the caffeine from the water. As in other methods, the caffeine can then be separated from the organic solvent by simple evaporation. The same water is recycled through this two-step process with new batches of beans. An equilibrium is reached after several cycles, wherein the water and the beans have a similar composition except for the caffeine. After this point, the caffeine is the only material removed from the beans, so no coffee strength or other flavorings are lost. Because water is used in the initial phase of this process, indirect method decaffeination is sometimes referred to as "water-processed".

Co_2 Process

This process has been referred to as CO_2 Method, Liquid Carbon Dioxide Method, and Supercritical Carbon Dioxide method but it is technically known as supercritical fluid extraction.

The supercritical CO_2 acts selectively on the caffeine, releasing the alkaloid and nothing else. Water-soaked coffee beans are placed in an extraction vessel. The extractor is then sealed and supercritical CO_2 is forced into the coffee at pressures of 1,000 pounds per square inch to extract the caffeine. The CO_2 acts as the solvent to dissolve and draw the caffeine from the coffee beans, leaving the larger-molecule flavor components behind. The caffeine-laden CO_2 is then transferred to another container called the absorption chamber where the pressure is released and the CO_2 returns to its gaseous state and evaporates, leaving the caffeine behind. The caffeine is removed from the CO_2 using charcoal filters, and the caffeine free CO_2 is pumped back into a pressurized container for reuse on another batch of beans. This process has the advantage that it avoids the use of potentially harmful substances. Because of its cost, this process is primarily used to decaffeinate large quantities of commercial-grade, less-exotic coffee found in grocery stores.

Triglyceride Process

Green coffee beans are soaked in a hot water/coffee solution to draw the caffeine to the surface of the beans. Next, the beans are transferred to another container and immersed in coffee oils that were obtained from spent coffee grounds and let to soak

After several hours of high temperatures, the triglycerides in the oils remove the caffeine, but not the flavor elements, from the beans. The beans are separated from the oils and dried. The caffeine is removed from the oils, which are reused to decaffeinate another batch of beans. This is a direct-contact method of decaffeination.

Decaffeinated Coffee

Caffeine Content of Coffee

Caffeine Content of Decaffeinated Coffee

To ensure product quality, manufacturers are required to test the newly decaffeinated coffee beans to make sure that caffeine concentration is relatively low (no less than 97% caffeine content reduction according to United States standards). To do so, many coffee companies choose to employ High-performance liquid chromatography to quantitatively measure how much caffeine remains in the coffee beans. However, since HPLC can be quite costly, some coffee companies are beginning to use other methods such as Near-infrared (NIR) spectroscopy. Although HPLC is highly accurate, NIR spectroscopy is much faster, cheaper and overall easier to use. Lastly, another method typically used to quantify remaining caffeine includes Ultraviolet–visible spectroscopy, which can be greatly advantageous for decaffeination processes that include supercritical CO_2, as CO_2 does not absorb in the UV-Vis range.

A controlled study of ten samples of prepared decaffeinated coffee from coffee shops showed that some caffeine remained. Fourteen to twenty cups of such decaffeinated coffee would contain as much caffeine as one cup of regular coffee. The 16-ounce (473-ml) cups of coffee samples contained caffeine in the range of 8.6 mg to 13.9 mg. In another study of popular brands of decaf coffees, the caffeine content varied from 3 mg to 32 mg. An 8-ounce (237-ml) cup of regular coffee contains 95–200 mg of caffeine, and a 12-ounce (355-milliliter) serving of Coca-Cola contains 36 mg.

Both of these studies tested the caffeine content of store-brewed coffee, suggesting that the caffeine may be residual from the normal coffee served rather than poorly decaffeinated coffee.

Decaffito

As of 2009, progress toward growing coffee beans that do not contain caffeine was still continuing. The term "Decaffito" has been coined to describe this type of decaffeinated coffee, and trademarked in Brazil.

The prospect for Decaffito-type coffees was shown by the discovery of the naturally caffeine-free *Coffea charrieriana*, reported in 2004. It has a deficient caffeine synthase gene, leading it to accumulate theobromine instead of converting it to caffeine. Either this trait could be bred into other coffee plants by crossing them with *C. charrieriana*, or an equivalent effect could be achieved by knocking out the gene for caffeine synthase in normal coffee plants.

Decaffeinated Tea

Tea may also be decaffeinated, usually by using processes analogous to the direct method or the CO_2 process, as described above. The process of oxidizing tea leaves to create black tea ("red" in Chinese tea culture) or oolong tea leaves from green leaves, does not affect the amount of caffeine in the tea, though tea-plant species (i.e., *Camellia sinensis sinensis* vs. *Camellia sinensis assamica*) may differ in natural caffeine content. Younger leaves and buds contain more caffeine per weight than older leaves and stems. Although the CO_2 process is favorable because it is convenient, nonexplosive, and nontoxic, a comparison between regular and decaffeinated green teas using supercritical carbon dioxide showed that most volatile, nonpolar compounds (such as linalool and phenylacetaldehyde), green and floral flavor compounds (such as hexanal and (*E*)-2-hexenal), and some unknown compounds disappeared or decreased after decaffeination.

In addition to CO_2 process extraction, tea may be also decaffeinated using a hot water treatment. Optimal conditions are met by controlling water temperature, extraction time, and ratio of leaf to water, where higher temperatures at or over 100 °C, moderate extraction time of 3 minutes, and a 1:20 water to leaf weight per volume ratio removed 83% caffeine content and preserved 95% of total catechins. Catechins, a type of flavanol, contribute to the flavor of the tea and also, interestingly, have been shown to increase suppression of mutagens that may lead to cancer.

Both coffee and tea have tannins, which is responsible for the astringent taste, but tea has nearly three times smaller tannin content than coffee. Thus, decaffeination of tea requires more care to maintain tannin content than decaffeination of coffee in order to preserve this flavor. Preserving tannins is desirable not only because of their flavor, but also because they have been shown to have anticarcinogenic, antimutagenic, antioxidative, and antimicrobrial properties. Specifically, tannins accelerate blood clotting, reduce blood pressure, decrease the serum lipid level, produce liver necrosis, and modulate immunoresponses.

Certain processes during normal production might help to decrease the caffeine content directly, or simply lower the rate at which it is released throughout each infusion. Several instances in China where this is evident is in many cooked pu-erh teas, as well as more heavily fired Wuyi Mountain oolongs; commonly referred to as 'zhonghuo' (mid-fired) or 'zuhuo' (high-fired).

A generally accepted statistic is that a cup of normal black (often called red in China; as distinct from green) tea contains 40–50 mg of caffeine, roughly half the content of a cup of coffee.

Although a common technique of discarding a short (30- to 60-second) steep is believed to much reduce caffeine content of a subsequent brew at the cost of some loss of flavor, research suggests that a five-minute steep yields up to 70% of the caffeine, and a second steep has one-third the caffeine of the first (about 23% of the total caffeine in the leaves).

Ultra-high-temperature Processing

Ultra-high temperature processing (UHT), ultra-heat treatment, or ultra-pasteurization sterilizes food by heating it above 135 °C (275 °F) – the temperature required to kill spores in milk – for 1 to 2 seconds. UHT is most commonly used in milk production, but the process is also used for fruit juices, cream, soy milk, yogurt, wine, soups, honey, and stews. UHT milk was first developed in the 1960s and became generally available for consumption in the 1970s.

The heat used during the UHT process can cause Maillard browning and change the taste and smell of dairy products. An alternative process is HTST pasteurization (high temperature/short time), in which the milk is heated to 72 °C (162 °F) for at least 15 seconds.

UHT milk packaged in a sterile container, if not opened, has a typical unrefrigerated shelf life of six to nine months. HTST pasteurized milk has a shelf life of about two weeks from processing, or about one week from being put on sale.

A Tetra Pak ultra-pasteurization line.

History

The most commonly applied technique to provide a safe and shelf-stable milk is heat treatment. The first system involving indirect heating with continuous flow (125 °C for 6 min) was manufactured in 1893. In 1912, a continuous-flow, direct-heating method of mixing steam with milk at temperatures of 130 to 140 °C was patented. However, without commercially available aseptic packaging systems to pack and store the product, such technology was not much useful in itself, and further development was stalled until the 1950s. In 1953, APV pioneered a steam injection technology, involving direct injection of steam through a specially designed nozzle which raises the product temperature instantly, under brand name Uperiser; milk was packaged in sterile cans. In the 1960s APV launched the first commercial steam infusion system under the Palarisator brand name.

In Sweden, Tetra Pak launched tetrahedral paperboard cartons in 1952. They made a commercial breakthrough in the 1960s, after technological advances, combining carton assembling and aseptic packaging technologies, followed by international expansion. In aseptic processing, the product and the package are sterilized separately and then combined and sealed in a sterile atmosphere, in contrast to canning, where product and package are first combined and then sterilized.

Technology

Ultra-high-temperature processing is performed in complex production plants, which perform several stages of food processing and packaging automatically and in succession:

- Heating
- Flash cooling
- Homogenization
- Aseptic packaging

In the heating stage, the treated liquid is first pre-heated to a non-critical temperature (70–80°C for milk), and then quickly heated to the temperature required by the process. There are two types of heating technologies: *direct*, where the product is put in a direct contact with the hot steam, and *indirect*, where the product and the heating medium remain separated by the equipment's contact surfaces. The main goals of the design, both from product quality and from efficiency standpoints, are to maintain the high product temperature for the shortest period possible, and to ensure that the temperature is evenly distributed throughout.

Direct Heating Systems

Direct systems have an advantage that the product is held at a high temperature for a shorter period of time, thereby reducing the thermal damage for the sensitive products such as milk. There are two groups of direct systems:

- Injection-based, where the high-pressure steam is injected into the liquid. It allows fast heating and cooling, but is only suitable for some products. As the product comes in contact with the hot nozzle, there is a possibility of local overheating.
- Infusion-based, where the liquid is pumped through a nozzle into a chamber with high-pressure steam at a relatively low concentration, providing a large surface contact area. This method achieves near-instantaneous heating and cooling and even distribution of temperature, avoiding local overheating. It is suitable for liquids of both low and high viscosity.

Indirect Heating Systems

In indirect systems, the product is heated by a solid heat exchanger similar to those used for pasteurization. However, as higher temperatures are applied, it is necessary to employ higher pressures in order to prevent boiling. There are three types of exchangers in use:

- Plate exchangers,
- Tubular exchangers
- Scraped-surface exchangers.

For higher efficiency, pressurized water or steam is used as the medium for heating the exchangers themselves, accompanied with a regeneration unit which allows reuse of the medium and energy saving.

Flash Cooling

After heating, the hot product is passed to a holding tube and then to a vacuum chamber, where it suddenly loses the temperature and vaporizes. The process, referred to as *flash cooling*, reduces the risk of thermal damage, removes some or all of excess water obtained through the contact with steam, and removes some of volatile compounds which negatively affect the product quality. Cooling rate and the quantity of water removed is determined by the level of vacuum, which must be carefully calibrated.

Homogenization

Homogenization is part of the process specific for milk. Homogenization is a mechanical treatment which results in decrease of size and increase of number and total surface area of fat globules in milk number. That, in turn, reduces milk's tendency to form cream at the surface and on contacts with container, enhances its stability, and makes it more palatable for consumers.

Aseptic Packaging

Aseptic packaging refers to technique in which previously sterilised milk is aseptically packaged in a sterile package and hermetically sealed to have prolonged shelf life even under ambient conditions.

Worldwide Use

UHT milk has seen large success in much of Europe, where across the continent as a whole 7 out of 10 Europeans drink it regularly. In fact, in a hot country such as Spain, UHT milk is preferred due to high costs of refrigerated transportation and "inefficient cool cabinets". UHT is less popular in Northern Europe and Scandinavia, particularly in Denmark, Finland, Norway, Sweden, the United Kingdom and Ireland. It is also less popular in Greece, where fresh pasteurized milk is the most popular type of milk.

UHT milk as a percentage of total consumption	
Country	**percent**
Austria	20.3
Belgium	96.7
Croatia	73
Czech Republic	71.4
Denmark	0.0
Finland	2.4
France	95.5
Brazil	88.1
Germany	66.1
Greece	0.9
Hungary	35.1
Ireland	10.9
Italy	49.8
Netherlands	20.2
Norway	5.3
Poland	48.6
Portugal	92.9
Slovakia	35.5
Spain	95.7
Sweden	5.5
Switzerland	62.8

Turkey	53.1
United Kingdom	8.4

In June 1993, Parmalat introduced its UHT milk to the United States. In the American market, consumers are uneasy about consuming milk that is not delivered under refrigeration, and reluctant to buy it. To combat this, Parmalat is selling its UHT milk in old-fashioned containers, unnecessarily sold from the refrigerator aisle. UHT milk is also used for many dairy products.

UHT milk is sold on American military bases in Puerto Rico and Korea due to limited availability of milk supplies and refrigeration.

UHT milk gained popularity in Puerto Rico as an alternative to pasteurized milk due to environmental factors. For example, power outages after a hurricane can last up to 2 weeks, during which time pasteurized milk would spoil from lack of refrigeration.

In 2008 the UK government proposed a 90% UHT milk production target by 2020 which they believed would significantly cut the need for refrigeration, and thus benefit the environment by reducing green house emissions. However the milk industry opposed this, and the proposition was abandoned.

Nutritive Effects

- Calories

  UHT milk contains the same number of calories as pasteurized milk.

- Calcium

  UHT and pasteurized milk contains the same amount of calcium.

- Protein

  UHT milk's protein structure is different from that of pasteurized milk, which prevents it from separating in cheese making.

- Folate

  UHT milk contains 1 µg of folate per 100 g, while pasteurized milk contains 9 µg.

- Vitamin B_{12}, vitamin C and thiamin

  Some nutritional loss can occur in UHT milk.

Cartons of UHT milk

Food Engineering

Food engineering is a multidisciplinary field of applied physical sciences which combines science, microbiology, and engineering education for food and related industries. Food engineering includes, but is not limited to, the application of agricultural engineering, mechanical engineering and chemical engineering principles to food materials. Food engineers provide the technological knowledge transfer essential to the cost-effective production and commercialization of food products and services. Physics, chemistry, and mathematics are fundamental to understanding and engineering products and operations in the food industry.

Food engineering encompasses a wide range of activities. Food engineers are employed in food processing, food machinery, packaging, ingredient manufacturing, instrumentation, and control. Firms that design and build food processing plants, consulting firms, government agencies, pharmaceutical companies, and health-care firms also employ food engineers. Specific food engineering activities include:

- drug/food products;
- design and installation of food/biological/pharmaceutical production processes;
- design and operation of environmentally responsible waste treatment systems;
- marketing and technical support for manufacturing plants.

Topics in Food Engineering

In the development of food engineering, one of the many challenges is to employ modern tools, technology, and knowledge, such as computational materials science and nanotechnology, to develop new products and processes. Simultaneously, improving quality, safety, and security remain critical issues in food engineering study. New packaging materials and techniques are being developed to provide more protection to foods, and novel preservation technology is emerging. Additionally, process control and automation regularly appear among the top priorities identified in food engineering. Advanced monitoring and control systems are developed to facilitate automation and flexible food manufacturing. Furthermore, energy saving and minimization of environmental problems continue to be important food engineering issues, and significant progress is being made in waste management, efficient utilization of energy, and reduction of effluents and emissions in food production.

Typical topics include:food people

- Advances in classical unit operations in engineering applied to food manufacturing
- Progresses in the transport and storage of liquid and solid foods
- Developments in heating, chilling and freezing of foods
- Advanced mass transfer in foods

- New chemical and biochemical aspects of food engineering and the use of kinetic analysis
- New techniques in dehydration, thermal processing, non-thermal processing, extrusion, liquid food concentration, membrane processes and applications of membranes in food processing
- Shelf-life, electronic indicators in inventory management, and sustainable technologies in food processing
- Modern packaging, cleaning, and sanitation technologies.
- Development of sensors systems for quality and safety assessment

Food Packaging

Food packaging is packaging for food. A package provides protection, tampering resistance, and special physical, chemical, or biological needs. It may bear a nutrition facts label and other information about food being offered for sale.

Testing modified atmosphere in a plastic bag of carrots

Functions

Packaging and package labeling have several objectives

- Physical protection - The food enclosed in the package may require protection from, among other things, shock, vibration, compression, temperature, bacteria, etc.
- Barrier protection - A barrier from oxygen, water vapor, dust, etc., is often required. Permeation is a critical factor in design. Some packages contain desiccants or oxygen absorbers to help extend shelf life. Modified atmospheres or controlled atmospheres are also maintained in some food packages. Keeping the contents clean, fresh, and safe for the intended shelf life is a primary function.

- Containment or agglomeration - Small items are typically grouped together in one package to allow efficient handling. Liquids, powders, and granular materials need containment.
- Information transmission - Packages and labels communicate how to use, transport, recycle, or dispose of the package or product. Some types of information are required by governments.
- Marketing - The packaging and labels can be used by marketers to encourage potential buyers to purchase the product. Package design has been an important and constantly evolving phenomenon for several decades. Marketing communications and graphic design are applied to the surface of the package and (in many cases) the point of sale display.
- Security - Packaging can play an important role in reducing the security risks of shipment. Packages can be made with improved tamper resistance to deter tampering and also can have tamper-evident features to help indicate tampering. Packages can be engineered to help reduce the risks of package pilferage; some package constructions are more resistant to pilferage and some have pilfer-indicating seals. Packages may include authentication seals to help indicate that the package and contents are not counterfeit. Packages also can include anti-theft devices, such as dye packs, RFID tags, or electronic article surveillance tags, that can be activated or detected by devices at exit points and require specialized tools to deactivate. Using packaging in this way is a means of retail loss prevention.
- Convenience - Packages can have features which add convenience in distribution, handling, stacking, display, sale, opening, reclosing, use, and reuse.
- Portion control - Single-serving packaging has a precise amount of contents to control usage. Bulk commodities (such as salt) can be divided into packages that are a more suitable size for individual households. It also aids the control of inventory: selling sealed one-liter bottles of milk, rather than having people bring their own bottles to fill themselves.

Types

The above materials are fashioned into different types of food packages and containers such as:

Packaging Type		
Aseptic processing	Primary	Liquid whole eggs or dairy products
Trays	Primary	Portion of fish or meat
Bags	Primary	Potato chips, apples, rice
Boxes	Secondary	Corrugated box of primary packages: box of cereal cartons, frozen pizzas
Cans	Primary	Can of tomato soup
Cartons	Primary	Carton of eggs, milk or juice cartons
Flexible packaging	Primary	Bagged salad
Pallets	Tertiary	A series of boxes on a single pallet used to transport from the manufacturing plant to a distribution center

Wrappers	Tertiary	Used to wrap the boxes on the pallet for transport

Primary packaging is the main package that holds the food that is being processed. Secondary packaging combines the primary packages into one box being made. Tertiary packaging combines all of the secondary packages into one pallet.

Packaging Machines

A choice of packaging machinery requires consideration of technical capabilities, labor requirements, worker safety, maintainability, serviceability, reliability, ability to integrate into the packaging line, capital cost, floorspace, flexibility (change-over, materials, etc.), energy usage, quality of outgoing packages, qualifications (for food, pharmaceuticals, etc.), throughput, efficiency, productivity, and ergonomics, at a minimum.

Shrink-wrapping trays of bakery goods

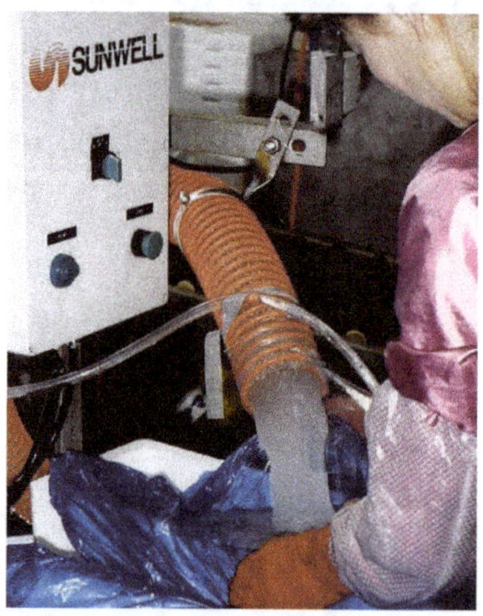

Pumping slurry ice onto fresh fish

Packaging machines may be of the following general types:

- Autocoding label and date verification
- Blister-, skin- and vacuum-packaging machines
- Capping, over-capping, lidding, closing, seaming and sealing machines
- Cartoning machines

- Case and tray forming, packing, unpacking, closing and sealing Machines
- Check weighing machines
- Cleaning, sterilizing, cooling and drying machines
- Conveying and accumulating machines
- Feeding, orienting, and placing machines
- Filling machines for liquid and powdered products
- Package filling and closing Machines
- Form, fill and seal machines
- Inspecting, detecting and checkweighing machines
- Palletizing, depalletizing, and pallet unitizing machines
- Labeling, marking, and other product identification machines
- Wrapping machines
- Converting machines

Reducing Food Packaging

Reduced packaging and sustainable packaging are becoming more frequent. The motivations can be government regulations, consumer pressure, retailer pressure, and cost control. Reduced packaging often saves packaging costs.

In the UK, a Local Government Association survey produced by the British Market Research Bureau compared a range of outlets to buy 29 common food items and found that small local retailers and market traders "produced less packaging and more that could be recycled than the larger supermarkets."

Recycling of Food Packaging

After use, organic matter that is still in the food packaging needs to be separated from the packaging. This may also require rinsing of the food packaging.

Trends in Food Packaging

- Numerous reports industry associations agree that use of smart indicators will increase. There are a number of different indicators with different benefits for food producers, consumers and retailers.
- Temperature recorders are used to monitor products shipped in a cold chain and to help validate the cold chain. Digital temperature data loggers measure and record the temperature history of food shipments. They sometimes have temperatures displayed on the indicator or have other output (lights, etc.): The data from a shipment can be downloaded (cable, RFID,

etc.) to a computer for further analysis. These help identify if there has been temperature abuse of products and can help determine the remaining shelf life. They can also help determine the time of temperature extremes during shipment so corrective measures can be taken.

    - Time temperature indicators integrate the time and temperature experienced by the indicator and adjacent foods. Some use chemical reactions that result in a color change while others use the migration of a dye through a filter media. To the degree that these physical changes in the indicator match the degradation rate of the food, the indicator can help indicate probable food degradation.

- Radio Frequency Identification is applied to food packages for supply chain control and has shown a significant benefit in allowing food producers and retailers create full real time visibility of their supply chain.
- Plastic packaging being used is usually non-biodegradable due to possible interactions with the food. Also, biodegradable polymers often require special composting conditions to properly degrade. Normal sealed landfill conditions do not promote biodegradation. Biodegradable plastics includes biodegradable films and coatings synthesized from organic materials and microbial polymers. Some package materials are edible. For example, pharmaceuticals are sometimes in capsules made of gelatin, starch, potato or other materials. Newer bioplastics, films and products are being developed.
- Barcodes have been used for decades in packaging many products. 2D barcodes used in Autocoding are increasingly applied to food packaging to ensure products are correctly packaged and date coded.
- The ability of a package to fully empty or dispense a viscous food is somewhat dependent on the surface energy of the inner walls of the container. The use of superhydrophobic surfaces is useful but can be further improved by using new lubricant-impregnated surfaces.

Food Safety and Public Health

It is critical to maintain food safety during processing , packaging, storage, logistics (including cold chain), sale, and use. Conformance to applicable regulations is mandatory. Some are country specific such as the US Food and Drug Administration and the US Department of Agriculture; others are regional such as the European Food Safety Authority. Certification programs such as the Global Food Safety Initiative are sometimes used. Food packaging considerations may include: use of hazard analysis and critical control points, verification and validation protocols, Good manufacturing practices, use of an effective quality management system, track and trace systems, and requirements for label content. Special food contact materials are used when the package is in direct contact with the food product. Depending on the packaging operation and the food, packaging machinery often needs specified daily washdown and cleaning procedures.

Health risks of materials and chemicals used in food packaging need to be carefully controlled. Carcinogens, toxic chemicals, mutagens etc. need to be eliminated from food contact and potential migration into foods.

Various Food Packing Machines

Autocoding

Autocoding refers to software solutions that help manufacturers, particularly those in the food industry, ensure that products have the correct packaging and correct 'sell by' date codes, thereby reducing the number of Emergency Product Withdrawals (EPW). The term was first used during an initiative between Geest PLC (acquired by Bakkavor in 2005) and Tesco PLC in 2001.

OAL Autocoding with Cognex

The key objective of the software was to reduce the number of EPW's associated with date and price coding errors, and pot and lid marriage errors. This still remains the main objective of autocoding software, but functionality has been expanded to encompass quality assurance and OEE performance data.

History

The concept of autocoding originally came from an initiative between Geest PLC and Tesco PLC in 2001, with both parties seeking to reduce the number of emergency product withdrawals associated with packaging and date coding errors. From here, a number of software engineering companies expanded on this preliminary work to create more reliable and robust systems. In 2004, 2D barcoding was introduced for the first time, a major step forward for the securing of all parts of food packaging.

Prior to this, 1D retail barcodes were used that only offered limited protection as they did not cover all packaging on products. 2D barcodes meant that lids, sleeves, pack could all be identified by a barcode scanner to check that the product is packaged and dated correctly. In 2009, Marks and Spencer introduced a code of practice for the labelling of products that stipulated an autocoding system must be used on every production line. In 2013, it was estimated that autocoding software protects over 1500 lines in the United Kingdom; the largest supplier is Olympus Automation, which protects 723 food production lines.

System Elements

All autocoding systems comprise a products database which contains standard reference information for each product including packaging type, labels and sell by date criteria. In most cases a touch screen industrial PC is positioned on the shop-floor to allow the operator to select the next product from a product schedule.

1D and 2D Barcode Scanning

The shop-floor touch screen device is linked to barcode scanners deployed to scan the code on each piece of packaging, including promotional labels and sleeves. Originally the bar codes scanned were based on standard 1D codes but to avoid mistakes 2D bar codes were introduced in 2004 so that each packaging type could hold a unique identity. To checks that the scanners are operational Autocoding solutions include two way communications with all hardware devices, or prevent the lines starting if links are not available.

Date Code Printing

To ensure that 'sell by dates' are accurate, most autocoding systems directly control the line printers through the software application. Once the operator has selected the product to run, the product reference table identifies the date range to use and the printer output is sent directly to the printer. Again, like the bar code scanners, autocoding systems include two-way communications with date code printers, and prevent the production lines from starting if links are not available.

Line Stops

If any error is detected, such as wrong film/pack, wrong lid, wrong case, or printer fault, the line is stopped. This is a standard requirement and is achieved through the use of PLCs.

OAL Autocoding system line stops

System Providers

Numerous companies provide autocoding solutions, ranging from standalone systems to comprehensive MES/MIS solutions that incorporate additional features and benefits. Notable industrial vendors include:

Cartoning Machine

A cartoning machine or cartoner, is a packaging machine that forms cartons: erect, close, folded, side seamed and sealed cartons.

Packaging machines which form a carton board blank into a carton filled with a product or bag of products or number of products say into single carton, after the filling, the machine engages its tabs / slots to apply adhesive and close both the ends of carton completely sealing the carton.

Cartoning machines can be divided into two types:

- Horizontal cartoning machines
- Vertical cartoning machines

A cartoning machine which picks a single piece from stack of folded carton and erects it, fills with a product or bag of products or number of products horizontally through an open end and closes by tucking the end flaps of the carton or applying glue or adhesive. The product might be pushed in the carton either through the mechanical sleeve or by pressurized air. For many applications however, the products are inserted into the carton manually. This type of Cartoning machine is widely used for packaging foodstuffs, confectionery, medicine, cosmetics, sundry goods, etc.

A cartoning machine which erects a folded carton, fills with a product or number of products vertically through an open end and closes by either tucking the end flaps of the carton or applying glue or adhesive, is called an end load cartoning machine. Cartoning machines are widely used for packaging bottled foodstuffs, confectionery, medicine, cosmetics, etc., and can vary based on the scale of business.

Check Weigher

A checkweigher is an automatic or manual machine for checking the weight of packaged commodities. It is normally found at the offgoing end of a production process and is used to ensure that the weight of a pack of the commodity is within specified limits. Any packs that are outside the tolerance are taken out of line automatically.

A checkweigher can weigh in excess of 500 items per minute (depending on carton size and accuracy requirements). Checkweighers can be used with metal detectors and X-ray machines to enable other attributes of the pack to be checked and acted upon accordingly.

A Typical Machine

An automatic checkweigher incorporates a series of conveyor belts. These checkweighers are known also as belt weighers, in-motion scales, conveyor scales, dynamic scales, and in-line scales. In filler applications, they are known as check scales. Typically, there are three belts or chain beds:

- An infeed belt that may change the speed of the package and to bring it up or down to a speed required for weighing. The infeed is also sometimes used as an indexer, which sets the gap between products to an optimal distance for weighing. It sometimes has special belts or chains to position the product for weighing.
- A weigh belt. This is typically mounted on a weight transducer which can typically be a strain-gauge load cell or a servo-balance (also known as a force-balance), or sometimes known as a split-beam. Some older machines may pause the weigh bed belt before taking the weight measurement. This may limit line speed and throughput.
- A reject belt that provides a method of removing an out-of-tolerance package from the conveyor line. The reject can vary by application. Some require an air-amplifier to blow small products off the belt, but heavier applications require a linear or radial actuator. Some fragile products are rejected by "dropping" the bed so that the product can slide gently into

a bin or other conveyor.

For high-speed precision scales, a load cell using electromagnetic force restoration (EMFR) is appropriate. This kind of system charges an inductive coil, effectively floating the weigh bed in an electromagnetic field. When the weight is added, the movement of a ferrous material through that coil causes a fluctuation in the coil current proportional to the weight of the object. Other technologies used include strain gauges and vibrating wire load cells.

It is usual for a built-in computer to take many weight readings from the transducer over the time that the package is on the weigh bed to ensure an accurate weight reading.

Calibration is critical. A lab scale, which usually is in an isolated chamber pressurized with dry nitrogen(pressurized at sea level) can weigh an object within plus or minus 100th of a gram, but ambient air pressure is a factor. This is straightforward when there is no motion, but in motion there is a factor that is not obvious-noise from the motion of a weigh belt, vibration, air-conditioning or refrigeration which can cause drafts. Torque on a load cell causes erratic readings.

A dynamic, in-motion checkweigher takes samples, and analyzes them to form an accurate weight over a given time period. In most cases, there is a trigger from an optical(or ultrasonic) device to signal the passing of a package. Once the trigger fires, there is a delay set to allow the package to move to the "sweet spot" (center) of the weigh bed to sample the weight. The weight is sampled for a given duration. If either of these times are wrong, the weight will be wrong. There seems to be no scientific method to predict these timings. Some systems have a "graphing" feature to do this, but it is generally more of an empirical method that works best.

- A reject conveyor to enable the out-of-tolerance packages to be removed from the normal flow while still moving at the conveyor velocity. The reject mechanism can be one of several types. Among these are a simple pneumatic pusher to push the reject pack sideways from the belt, a diverting arm to sweep the pack sideways and a reject belt that lowers or lifts to divert the pack vertically. A typical checkweigher usually has a bin to collect the out-of-tolerance packs. Sometimes these bins are provided with a lock, to prevent that out of specification items are fed back on the conveyor belt.

Tolerance Methods

There are several tolerance methods:

- The traditional "minimum weight" system where weights below a specified weight are rejected. Normally the minimum weight is the weight that is printed on the pack or a weight level that exceeds that to allow for weight losses after production such as evaporation of commodities that have a moisture content. The larger wholesale companies have mandated that any product shipped to them have accurate weight checks such that a customer can be confident that they are getting the amount of product for which they paid. These wholesalers charge large fees for inaccurately filled packages.
- The European Average Weight System which follows three specified rules known as the "Packers Rules".
- Other published standards and regulations such as NIST Handbook 133

Data Collection

There is also a requirement under the European Average Weight System that data collected by checkweighers is archived and is available for inspection. Most modern checkweighers are therefore equipped with communications ports to enable the actual pack weights and derived data to be uploaded to a host computer. This data can also be used for management information enabling processes to be fine-tuned and production performance monitored.

Checkweighers that are equipped with high speed communications such as Ethernet ports are capable of integrating themselves into groups such that a group of production lines that are producing identical products can be considered as one production line for the purposes of weight control. For example, a line that is running with a low average weight can be complemented by another that is running with a high average weight such that the aggregate of the two lines will still comply with rules.

An alternative is to program the checkweigher to check bands of different weight tolerances. For instance, the total valid weight is 100 grams ±15 grams. This means that the product can weigh 85 g - 115 g. However, it is obvious that if you are producing 10,000 packs a day, and most of your packs are 110 g, you are losing 100 kg of product. If you try to run closer to 85 g, you may have a high rejection rate.

EXAMPLE: A checkweigher is programmed to indicate 5 zones with resolution to 1 g:

1. Under Reject.... the product weighs 84.9 g or less
2. Under OK........ the product weighs 85 g, but less than 95 g
3. Valid........... the product weighs 96 g, but less than 105 g
4. Over OK......... the product weighs 105 g, and less than 114 g
5. Over Reject..... the product weighs over the 115 g limit

With a check weigher programmed as a zone checkweigher, the data collection over the networks, as well as local statistics, can indicate the need to check the settings on the upstream equipment to better control flow into the packaging. In some cases the dynamic scale sends a signal to a filler, for instance, in real-time, controlling the actual flow into a barrel, can, bag, etc. In many cases a checkweigher has a light-tree with different lights to indicate the variation of the zone weight of each product.

This data can be used to determine if an issue exists with an upstream filling, or packaging, machine. A checkweigher can send a signal to the machine to increase or decrease the amount put into a package. This can result in a payback associated with the checkweigher since producers will be better able to control the amount of give-away. See checkweigher case study outlining ground beef and packaging savings.

Application Considerations

Speed and accuracy that can be achieved by a checkweigher is influenced by the following:

- Pack length

- Pack weight
- Line speed required
- Pack content (solid or liquid)
- Motor technology
- Stabilization time of the weight transducer
- Airflow causing readings in error
- Vibrations from machinery causing unnecessary rejects
- Sensitivity to temperature, as the load cells *can* be temperature sensitive

Applications

In-motion scales are dynamic machines that can be designed to perform thousands of tasks. Some are used as simple caseweighers at the end of the conveyor line to ensure the overall finished package product is within its target weight.

An in motion conveyor checkweigher can be used to detect missing pieces of a kit, such as a cell phone package that is missing the manual, or other collateral. Checkweighers are typically used on the incoming conveyor chain, and the output pre-packaging conveyor chain in a poultry processing plant. The bird is weighed when it comes onto the conveyor, then after processing and washing at the end, the network computer can determine whether or not the bird absorbed too much water, which as it is further processed, will be drained, making the bird under its target weight.

A high speed conveyor scale can be used to change the pacing, or pitch of the products on the line by speeding, or slowing the product speed to change the distance between packs before reaching a different speed going into a conveyor machine that is boxing multiple packs into a box. The "pitch" is the measurement of the product as it comes down the conveyor line from leading edge to leading edge.

A checkweigher can be used to count packs, and the aggregate (total) weight of the boxes going onto a pallet for shipment, including the ability to read each package's weight and cubic dimensions. The controller computer can print a shipping label and a bar-code label to identify the weight, the cubic dimensions, ship-to address, and other data for machine ID through the shipment of the product. A receiving checkweigher for the shipment can read the label with a bar code scanner, and determine if the shipment is as it was before the transportation carrier received it from the shipper's loading dock, and determine if a box is missing, or something was pilfered or broken in transit.

Checkweighers are also used for Quality management. For instance, raw material for machining a bearing is weighed prior to beginning the process, and after the process, the quality inspector expects that a certain amount of metal was removed in the finishing process. The finished bearings are checkweighed, and bearings over- or underweight are rejected for physical inspection. This is a benefit to the inspector, since he can have a high confidence that the ones not rejected are within machining tolerance. A common usage is for throttling plastic extruders such that a bottle used to package detergent meets that requirements of the finished packager.

Quality management can use a checkweigher for Nondestructive testing to verify finished goods using common Evaluation methods to detect pieces missing from a "finished" product, such as grease from a bearing, or a missing roller within the housing.

Checkweighers can be built with metal detectors, x-ray machines, open-flap detection, bar-code scanners, holographic scanners, temperature sensors, vision inspectors, timing screws to set the timing and spacing between product, indexing gates and concentrator ducts to line up the product into a designated area on the conveyor. An industrial motion checkweigher can sort products from a fraction of a gram to many, many kilograms. In English units, is this from less than 100th of an ounce to as much as 500 lbs or more. Specialized checkweighers can weigh commercial aircraft, and even find their center-of-gravity.

Checkweighers can operate at very high speeds, processing products weighing fractions of a gram at over 100m/m (meters per minute) and materials such as pharmaceuticals and 200 lb bags of produce at over 100fpm(feet per minute). They can be designed in many shapes and sizes, hung from ceilings, raised on mezzanines, operated in ovens or in refrigerators. Their conveying medium can be industrial belting, low-static belting, chains similar to bicycle chains(but much smaller), or interlocked chain belts of any width. They can have chain belts made of special materials, different polymers, metals, etc.

Checkweighers are used in cleanrooms, dry atmosphere environments, wet environments, produce barns, food processing, drug processing, etc. Checkweighers are specified by the kind of environment, and the kind of cleaning will be used. Typically, a checkweigher for produce is made of mild steel, and one that will be cleaned with harsh chemicals, such as bleach, will be made with all stainless steel parts, even the Load cells. These machines are labeled "full washdown", and must have every part and component specified to survive the washdown environment.

Checkweighers are operated in some applications for extremely long periods of time- 24/7 year round. Generally, conveyor lines are not stopped unless there is maintenance required, or there is an emergency stop, called an E-stop. Checkweighers operating in high density conveyor lines may have numerous special equipments in their design to ensure that if an E-stop occurs, all power going to all motors is removed until the E-stop is cleared and reset.

Active Packaging

The terms active packaging, intelligent packaging, and smart packaging refer to packaging systems used with foods, pharmaceuticals, and several other types of products. They help extend shelf life, monitor freshness, display information on quality, improve safety, and improve convenience.

The terms are closely related. *Active packaging* usually means having active functions beyond the inert passive containment and protection of the product. *Intelligent* and *smart* packaging usually involve the ability to sense or measure an attribute of the product, the inner atmosphere of the package, or the shipping environment. This information can be communicated to users or can trigger active packaging functions. Programmable matter, smart materials, etc can be employed in packages.

Depending on the working definitions, some traditional types of packaging might be considered as "active" or "intelligent". More often, the terms are used with new technologically advanced systems: microelectronics, computer applications, nanotechnology, etc.

Moisture Control

For many years, desiccants have been used to actively control the water vapor in a closed package. A desiccant is a hygroscopic substance usually in a porous pouch or sachet which is placed inside a sealed package. They have been used to reduce corrosion of machinery and electronics and to extend the shelf life of moisture sensitive foods and drugs.

Corrosion

Corrosion inhibitors can be applied to items to help prevent rust and corrosion. Volatile corrosion inhibitors (VCI) or vapor phase corrosion inhibitors can be provided inside a package in a pouch or can be incorporated in a saturated overwrap of special paper. Many of these are organic salts that condense on the metal to resist corrosion. Some films also have VCI emitting capability.

Films are available with copper ions in the polymer structure, These neutralize the corrosive gas in a package and deter rust.

VCI create a neutral environment in the packaging. It works on the principle of difference in vapour pressure and causes reaction with Metals and Non metals and also with Moisture to render it active to corrosion. There are different forms of VCI's available like Papers, Plastics, HDPE Papers, Oils, Foams, Chips, Aluminum Barrier Foils, Bubble, Emitters etc. that can prevent corrosion at many stages.

Oxygen Control

Oxygen scavengers or oxygen absorbers help remove oxygen from a closed package. Some are small packets or sachets containing powdered iron: as the iron rusts, oxygen is removed from the surrounding atmosphere. Newer systems are on cards or can be built into package films or molded structures. In addition, the physical characteristics of the packaging itself (oxygen transmission rate - OTR) can dictate how effective an oxygen absorber can be, and how long it will stay effective. Packaging with a low OTR will let less oxygen in the closed package through the polymer barrier itself.

Atmosphere

With some products, such as cheese, it has long been common to flush the package with nitrogen prior to sealing: the inert nitrogen is absorbed into the cheese, allowing a tight shrink film package. The nitrogen removes oxygen and "actively" interacts with the cheese to make the package functional.

More recently, other mixtures of gas have been used inside the package to extend the shelf life. The gas mixture depends on the specific product and its degradation mechanisms. Some package components have been developed that incorporate active chemistry to help maintain certain atmospheres in packages.

Oxygen scavengers, carbon dioxide generators, ethanol generators, etc. are available to help keep the atmosphere in a package at specified conditions.

Temperature Monitor

Some temperature indicators give a visual signal that a specified temperature has been exceeded. Others, Time temperature indicators, signal when a critical accumulation of temperature deviation over time has been exceeded. When the mechanism of the indicator is tuned to the mechanism of product degradation, these can provide valuable signals for consumers.

Digital temperature data loggers record the temperatures encountered throughout the shipment. This data can be used to predict product degradation and help determine if the product is suited for normal sale or if expedited sale is required. They also determine the time of the temperature excess: this can be used to direct corrective action.

Thermochromic inks are sometimes used to signal temperature excess or change. Some are reversible while others have a permanent change of color. These can be used alone or with other packaging functions such as barcodes.

The inks can also signal a desired temperature for consumers. For example, one type of beer can has ink that graphically shows when an ideal drinking temperature is achieved.

Controlling Package Temperatures

For critical vaccines, insulated shipping containers are passive packaging to help control the temperatures fluctuations seen even with a controlled cold chain. In addition, gel packs are often used to actively keep the temperature of the contents within specified acceptable temperature ranges.

Some newer packages have the ability to heat or cool the product for the consumer. These have segregated compartments where exothermic or endothermic reactions provide the desired effect. Self-heating cans are available for several products.

Self-heating field rations for up to 18 soldiers

Dispensing

Some packages have closures or other dispensing systems that actively change the contents from a liquid to an aerosol. These are used for products ranging from precision inhalers for medications to spray bottles of household cleaners.

Some dispensing packages for two-part epoxy adhesives do more than passively contain the two components. When dispensed, some packages meter and mix the two components so the adhesive is fully functioning at the point of application.

The ability of a package to fully empty or dispense a viscous liquid is somewhat dependent on the surface energy of the inner walls of the container. The use of superhydrophobic surfaces is useful but can be further improved by using new lubricant-impregnated surfaces.

RFID

Radio-frequency identification chips are becoming more common as smart labels that are used to track and trace packages and unit loads throughout distribution. Newer developments include recording the temperature history of shipments and other intelligent packaging functions.

Security

A variety of security printing methods, security holograms, and specialized labels are available to help confirm that the product in the package is not counterfeit. RFID chips are being used in this application also.

Electronic article surveillance (on the product or on the package) is used to help counter shoplifting.

Microwave Packaging

Metallised films are used as a susceptor for cooking in microwave ovens. These increase the heating capacity and help make foods crisp and brown. Plastic microwavable containers are also used for microwave cooking.

Shock and Vibration

Shock detectors have been available for many years. These are attached to the package or to the product in the package to determine if an excessive shock has been encountered. The mechanisms of these *shock overload* devices have been spring-mass systems, magnets, drops of red dye, and several others.

Recently, digital shock and vibration data loggers have been available to more accurately record the shocks and vibrations of shipment. These are used to monitor critical shipments to determine if extra inspection and calibration is required. They are also used to monitor the types of shocks and vibrations encountered in transit for use in package testing in a laboratory.

Other Developments

Some engineered packaging films contain enzymes, anti-bacterial agents, scavengers, and other active components to help control food degradation and extend shelf life.

Edible films have been developed to allow consumers to eat the package along with the product.

Special packaging has been developed for shipping organs which keeps them alive during extended shipments. The organs are alive and fresh for transplanting.

Regulations

Active packaging is often designed to actively interact with the contents of the package. Thus extra care is often needed for active or smart packagings that are food contact materials.

Food packagers take extra care with some types of active packaging. For example, when the oxygen atmosphere in a package is reduced for extending shelf life, controls for anaerobic bacteria need to be considered. Also when a controlled atmosphere reduces the appearance of food degradation, consumers need to retain a means of determining whether actual degradation is present.

References

- Hans-Jürgen Bässler und Frank Lehmann : Containment Technology: Progress in the Pharmaceutical and Food Processing Industry. Springer, Berlin 2013, ISBN 978-3642392917
- Wayne Gisslen (2009). Professional Baking (5th ed.). Hoboken, N.J.: John Wiley. p. 77. ISBN 0-471-78349-8. Retrieved May 18, 2011.
- p 655 in "Advanced Dairy Chemistry: Volume 2 - Lipids" by P.F. Fox and P. McSweeney, Birkhäuser, 2006 ISBN 978-0-387-26364-9
- p 82 in "Lab Ref, Volume 2: A Handbook of Recipes, Reagents, and Other Reference Tools for Use at the Bench" by A.S. Mellick and L. Rodgers, CSHL Press, 2002, ISBN 978-0-87969-630-6
- Moyler, D (1993). "Extraction of flavours and fragrances with compressed CO2". In King, M; Bott, Theodore. Extraction of Natural Products Using Near-Critical Solvents (First ed.). Glasgow: Springer Netherlands. pp. 140–183. ISBN 978-94-010-4947-4.
- Alan H. Varnam (6 December 2012). Milk and Milk Products: Technology, chemistry and microbiology. Springer Science & Business Media. pp. 58–62. ISBN 978-1-4615-1813-6.
- Cichello, Simon. "A Guide to Oxygen Absorbers" (PDF). Archived from the original (PDF) on 27 August 2010. Retrieved 12 September 2016.
- Smith, J D; Dhiman, Rajeev; Anand, Sushant; Reza-Garduno, Ernesto; Cohen, Robert E.; McKinley, Gareth H.; Varanasi, Kripa K. (2013). "Droplet mobility on lubricant-impregnated surfaces". Soft Matter. 9 (6): 1772–1780. doi:10.1039/c2sm27032c. Retrieved 22 March 2015.
- International Agency for Research on Cancer. "Chemical agents and related occupations, Volume 100F. A review of human carcinogens." (PDF). International Agency for Research on Cancer. Retrieved 2014-08-20.
- Dainelli, D; Nathalie Gontard; Dimitrios Spyropoulos; Esther Zondervan-van den Beuken; Paul Tobback (2008). "Active and intelligent food packaging: legal aspects and safety concerns". Trends in Food Science & Technology. 19 (1): 167–177. doi:10.1016/j.tifs.2008.09.011. Retrieved 12 Nov 2014.
- В Рязани производят кошерное сухое молоко [Kosher Milk is Produced in Ryazan]. Наша Рязань (in Russian). 16 February 2010. Retrieved 22 April 2013.

9

Food Security: An Analytical Aspect

Food security refers to the storage of food to make its supply continuous and uninterrupted. Food security is a major concern and efforts are being made to ensure every person has access to food at all times. This chapter explores the essential aspects of food security, food safety, food grading and optical sorting. It is a major part of the food industry.

Food Security

Food security is a condition related to the supply of food, and individuals' access to it. Concerns over food security have existed throughout history. There is evidence of granaries being in use over 10,000 years ago, with central authorities in civilizations including ancient China and ancient Egypt being known to release food from storage in times of famine. At the 1974 World Food Conference the term "food security" was defined with an emphasis on supply. Food security, they said, is the "availability at all times of adequate world food supplies of basic foodstuffs to sustain a steady expansion of food consumption and to offset fluctuations in production and prices". Later definitions added demand and access issues to the definition. The final report of the 1996 World Food Summit states that food security "exists when all people, at all times, have physical and economic access to sufficient, safe and nutritious food to meet their dietary needs and food preferences for an active and healthy life".

Household food security exists when all members, at all times, have access to enough food for an active, healthy life. Individuals who are food secure do not live in hunger or fear of starvation. Food insecurity, on the other hand, is a situation of "limited or uncertain availability of nutritionally adequate and safe foods or limited or uncertain ability to acquire acceptable foods in socially acceptable ways", according to the United States Department of Agriculture (USDA). Food security incorporates a measure of resilience to future disruption or unavailability of critical food supply due to various risk factors including droughts, shipping disruptions, fuel shortages, economic instability, and wars. In the years 2011-2013, an estimated 842 million people were suffering from chronic hunger. The Food and Agriculture Organization of the United Nations, or FAO, identified the four pillars of food security as availability, access, utilization, and stability. The United Nations (UN) recognized the Right to Food in the Declaration of Human Rights in 1948, and has since noted that it is vital for the enjoyment of all other rights.

The 1996 World Summit on Food Security declared that "food should not be used as an instrument for political and economic pressure". According to the International Centre for Trade and Sustainable Development, failed agriculture market regulation and the lack of anti-dumping mechanisms cause much of the world's food scarcity and malnutrition.

Measurement

Food security indicators and measures are derived from country level household income and expenditure surveys to estimate per capita caloric availability. In general the objective of food security indicators and measures is to capture some or all of the main components of food security in terms of food availability, access and utilization or adequacy. While availability (production and supply) and utilization/adequacy (nutritional status/anthropometric measures) seemed much easier to estimate, thus more popular, access (ability to acquire sufficient quantity and quality) remain largely elusive. The factors influencing household food access are often context specific. Thus the financial and technical demands of collecting and analyzing data on all aspects of household's experience of food access and the development of valid and clear measures remain a huge challenge. Nevertheless, several measures have been developed that aim to capture the access component of food security, with some notable examples developed by the USAID-funded Food and Nutrition Technical Assistance (FANTA) project, collaborating with Cornell and Tufts University and Africare and World Vision. These include:

- *Household Food Insecurity Access Scale* (HFIAS) - continuous measure of the degree of food insecurity (access) in the household in the previous month
- *Household Dietary Diversity Scale* (HDDS) - measures the number of different food groups consumed over a specific reference period (24hrs/48hrs/7days).
- *Household Hunger Scale* (HHS)- measures the experience of household food deprivation based on a set of predictable reactions, captured through a survey and summarized in a scale.
- *Coping Strategies Index* (CSI) - assesses household behaviours and rates them based on a set of varied established behaviours on how households cope with food shortages. The methodology for this research is base on collecting data on a single question: "What do you do when you do not have enough food, and do not have enough money to buy food?"

Food insecurity is measured in the United States by questions in the Census Bureau's Current Population Survey. The questions asked are about anxiety that the household budget is inadequate to buy enough food, inadequacy in the quantity or quality of food eaten by adults and children in the household, and instances of reduced food intake or consequences of reduced food intake for adults and for children. A National Academy of Sciences study commissioned by the USDA criticized this measurement and the relationship of "food security" to hunger, adding "it is not clear whether hunger is appropriately identified as the extreme end of the food security scale."

The FAO, World Food Programme (WFP), and International Fund for Agricultural Development (IFAD) collaborate to produce *The State of Food Insecurity in the World*. The 2012 edition described improvements made by the FAO to the prevalence of undernourishment (PoU) indicator that is used to measure rates of food insecurity. New features include revised minimum dietary energy requirements for individual countries, updates to the world population data, and estimates of food losses in retail distribution for each country. Measurements that factor into the indicator include dietary energy supply, food production, food prices, food expenditures, and volatility of the food system. The stages of food insecurity range from food secure situations to full-scale famine. A new peer-reviewed journal of *Food Security: The Science, Sociology and Economics of Food Production and Access to Food* began publishing in 2009.

Rates

With its prevalence of undernourishment (PoU) indicator, the FAO reported that almost 870 million people were chronically undernourished in the years 2010-2012. This represents 12.5% of the global population, or 1 in 8 people. Higher rates occur in developing countries, where 852 million people (about 15% of the population) are chronically undernourished. The report noted that Asia and Latin America have achieved reductions in rates of undernourishment that put these regions on track for achieving the Millennium Development Goal of halving the prevalence of undernourishment by 2015. The UN noted that about 2 billion people do not consume a sufficient amount of vitamins and minerals. In India, the second-most populous country in the world, 30 million people have been added to the ranks of the hungry since the mid-1990s and 46% of children are underweight.

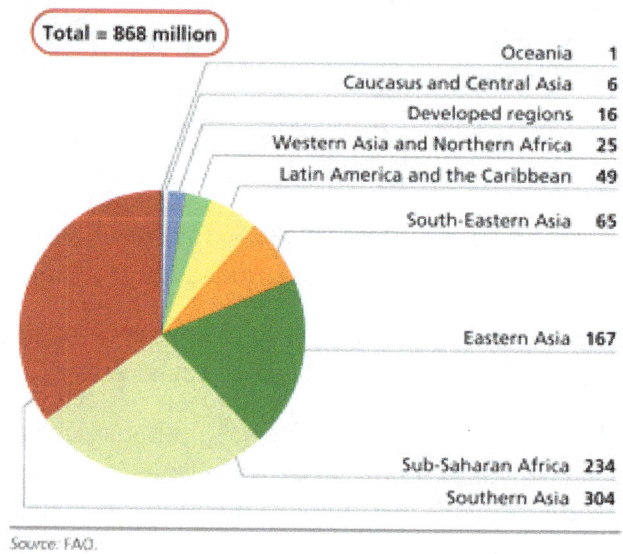

Number of people affected by undernourishment, 2010–12 (by region, in millions). Data source: FAO.

History

2007-08 Crisis

In late 2007, export restrictions and panic buying, US dollar depreciation, increased farming for use in biofuels, world oil prices at more than $100 a barrel, global population growth, climate change, loss of agricultural land to residential and industrial development, and growing consumer demand in China and India are claimed to have pushed up the price of grain. However, the role of some of these factors is under debate. Some argue the role of biofuel has been overplayed as grain prices have come down to the levels of 2006. Nonetheless, food riots took place in many countries across the world in 2011.

Food Security in the US

The United States Department of Agriculture defines food insecurity as "limited or uncertain availability of nutritionally adequate and safe foods or limited or uncertain ability to acquire acceptable foods in socially acceptable ways."

Prevalence of food insecurity, 2012

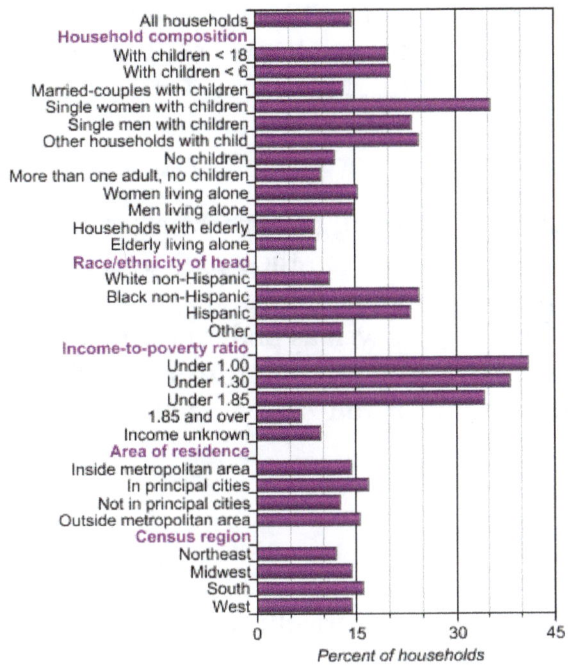

Source: Calculated by ERS using data from the December 2012 Current Population Survey Food Security Supplement.

National Food Security Surveys are the main survey tool used by the USDA to measure food security in the United States. Based on respondents' answers to survey questions, the household can be placed on a continuum of food security defined by the USDA. This continuum has four categories: high food security, marginal food security, low food security, and very low food security. Economic Research Service report number 155 (ERS-155) estimates that 14.5 percent (17.6 million) of US households were food insecure at some point in 2012. The prevalence of food insecurity has been relatively in the United States since the economic recession 2008.

In 2012:

- 49.0 million people lived in food-insecure households.
- 12.4 million adults lived in households with very low food security.
- 8.3 million children lived in food-insecure households in which children, along with adults, were food insecure.
- 977,000 children lived in households in which one or more child experienced very low food security.

Feed The Future

The government of the United States has taken up an initiative along with other local government agencies as well as global partners, the G8 countries, and donors to reduce global hunger and to improve food security conditions in the world. Exploiting the G8 Summit of 2009 held in L'Aquila, Italy, President Barack Obama encouraged global leaders to revert the three-decade old trend of reduced agricultural investments, and instead, choose to increase cash flow in their respective agricultural sectors in a drive to bolster global food security. This led to the birth of the "Feed the Future" program.

Initially, the U.S had been successful in collecting over $18 billion in funds for the program from the G8 countries as well as other donors. The "Feed the Future" initiative is presently led by the U.S Agency for International Development and has the support of other government-funded bodies like the State Department, Peace Corps, Millennium Challenge Corporation, the Treasury Department, U.S. Trade Representative, Overseas Private Investment Corporation, the U.S. African Development Foundation, and the U.S Department of Agriculture.

The target group of this initiative is the small holder farmers, especially women. Feed the Future has plans to help partner countries to develop their agricultural sector with an aim to improve their agricultural output. A spurt in economic growth would ultimately lead to higher income and help to eradicate hunger, and poverty, as well as under nutrition from the society. This initiative is expected to work on the basis of country-led priorities that call for consistent support by the governments, donor organizations, the private sector, and the civil society to accomplish its long-term goals.

World Summit on Food Security

The World Summit on Food Security held in Rome in 1996, aimed to renew a global commitment to the fight against hunger. The Food and Agriculture Organization of the United Nations (FAO) called the summit in response to widespread under-nutrition and growing concern about the capacity of agriculture to meet future food needs. The conference produced two key documents, the Rome Declaration on World Food Security and the World Food Summit Plan of Action.

The Rome Declaration calls for the members of the United Nations to work to halve the number of chronically undernourished people on the Earth by the year 2015. The Plan of Action sets a number of targets for government and non-governmental organizations for achieving food security, at the individual, household, national, regional and global levels.

Another World Summit on Food Security took place in Rome between November 16 and 18, 2009. The decision to convene the summit was taken by the Council of FAO in June 2009, at the proposal of FAO Director-General Dr Jacques Diouf. Heads of State and Government attended the summit, which took place at the FAO's headquarters.

Pillars of Food Security

The WHO states that there are three pillars that determine food security: food availability, food access, and food use. The FAO adds a fourth pillar: the stability of the first three dimensions of food security over time. In 2009, the World Summit on Food Security stated that the "four pillars of food security are availability, access, utilization, and stability".

Availability

Food availability relates to the supply of food through production, distribution, and exchange. Food production is determined by a variety of factors including land ownership and use; soil management; crop selection, breeding, and management; livestock breeding and management; and harvesting. Crop production can be affected by changes in rainfall and temperatures. The use of land, water, and energy to grow food often competes with other uses, which can affect food production. Land used for agriculture can be used for urbanization or lost to desertification, salinization, and soil erosion due to unsustainable agricultural practices. Crop production is not required for a country to achieve food security. Nations don't have to have the natural resources required to produce crops in order to achieve food security, as seen in the examples of Japan and Singapore.

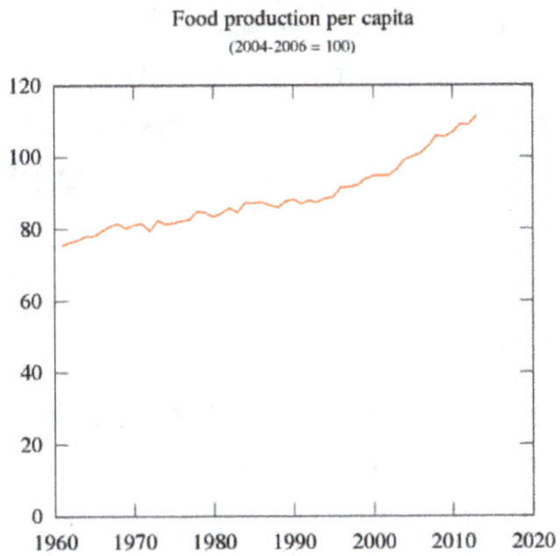

Growth in food production has been greater than population growth. Food per person increased since 1961. Data source: Food and Agriculture Organization.

Because food consumers outnumber producers in every country, food must be distributed to different regions or nations. Food distribution involves the storage, processing, transport, packaging, and marketing of food. Food-chain infrastructure and storage technologies on farms can also affect the amount of food wasted in the distribution process. Poor transport infrastructure can increase the price of supplying water and fertilizer as well as the price of moving food to national and global markets. Around the world, few individuals or households are continuously self-reliant for food. This creates the need for a bartering, exchange, or cash economy to acquire food. The exchange of food requires efficient trading systems and market institutions, which can affect food security. Per capita world food supplies are more than adequate to provide food security to all, and thus food accessibility is a greater barrier to achieving food security.

Access

Goats are an important part of the solution to global food security because they are fairly low-maintenance and easy to raise and farm.

Food access refers to the affordability and allocation of food, as well as the preferences of individuals and households. The UN Committee on Economic, Social, and Cultural Rights noted that the causes of hunger and malnutrition are often not a scarcity of food but an inability to access available food, usually due to poverty. Poverty can limit access to food, and can also increase how vulnerable an individual or household is to food price spikes. Access depends on whether the household has enough income to purchase food at prevailing prices or has sufficient land and other resources to grow its own food. Households with enough resources can overcome unstable harvests and local food shortages and maintain their access to food.

There are two distinct types of access to food: direct access, in which a household produces food using human and material resources, and economic access, in which a household purchases food produced elsewhere. Location can affect access to food and which type of access a family will rely on. The assets of a household, including income, land, products of labor, inheritances, and gifts can determine a household's access to food. However, the ability to access to sufficient food may not lead to the purchase of food over other materials and services. Demographics and education levels of members of the household as well as the gender of the household head determine the preferences of the household, which influences the type of food that are purchased. A household's access to enough and nutritious food may not assure adequate food intake of all household members, as intrahousehold food allocation may not sufficiently meet the requirements of each member of the household. The USDA adds that access to food must be available in socially acceptable ways, without, for example, resorting to emergency food supplies, scavenging, stealing, or other coping strategies.

Utilization

The next pillar of food security is food utilization, which refers to the metabolism of food by individuals. Once food is obtained by a household, a variety of factors affect the quantity and quality of food that reaches members of the household. In order to achieve food security, the food ingested must be safe and must be enough to meet the physiological requirements of each individual. Food safety affects food utilization, and can be affected by the preparation, processing, and cooking of food in the community and household. Nutritional values of the household determine food choice, and whether food meets cultural preferences is important to utilization in terms of psychological and social well-being. Access to healthcare is another determinant of food utilization, since the health of individuals controls how the food is metabolized. For example, intestinal parasites can take nutrients from the body and decrease food utilization. Sanitation can also decrease the occurrence and spread of diseases that can affect food utilization. Education about nutrition and food preparation can affect food utilization and improve this pillar of food security. education bring accountability among people,create awareness,which foster them for food security

Stability

Food stability refers to the ability to obtain food over time. Food insecurity can be transitory, seasonal, or chronic. In transitory food insecurity, food may be unavailable during certain periods of time. At the food production level, natural disasters and drought result in crop failure and decreased food availability. Civil conflicts can also decrease access to food. Instability in markets resulting in food-price spikes can cause transitory food insecurity. Other factors that can temporarily cause food insecurity are loss of employment or productivity, which can be caused by illness. Seasonal food insecurity can result from the regular pattern of growing seasons in food production.

Chronic (or permanent) food insecurity is defined as the long-term, persistent lack of adequate food. In this case, households are constantly at risk of being unable to acquire food to meet the needs of all members. Chronic and transitory food insecurity are linked, since the reoccurrence of transitory food security can make households more vulnerable to chronic food insecurity.

Effects of Food Insecurity

"Famine and hunger are both rooted in food insecurity. Chronic food insecurity translates into a high degree of vulnerability to famine and hunger; ensuring food security presupposes elimination of that vulnerability."

Stunting and Chronic Nutritional Deficiencies

Many countries experience ongoing food shortages and distribution problems. These result in chronic and often widespread hunger amongst significant numbers of people. Human populations can respond to chronic hunger and malnutrition by decreasing body size, known in medical terms as stunting or stunted growth. This process starts *in utero* if the mother is malnourished and continues through approximately the third year of life. It leads to higher infant and child mortality, but at rates far lower than during famines. Once stunting has occurred, improved nutritional intake after the age of about two years is unable to reverse the damage. Stunting itself can be viewed as a coping mechanism, bringing body size into alignment with the calories available during adulthood in the location where the child is born. Limiting body size as a way of adapting to low levels of energy (calories) adversely affects health in three ways:

Children with symptoms of low calorie and protein intake and a nurse attendant at a Nigerian orphanage in the late 1960s.

- Premature failure of vital organs during adulthood. For example, a 50-year-old individual might die of heart failure because his/her heart suffered structural defects during early development;
- Stunted individuals suffer a higher rate of disease and illness than those who have not undergone stunting;
- Severe malnutrition in early childhood often leads to defects in cognitive development. It therefore creates disparity among children who did not experience severe malnutrition and those who did.

Challenges to Achieving Food Security

Global Water Crisis

Water deficits, which are already spurring heavy grain imports in numerous smaller countries, may soon do the same in larger countries, such as China or India. The water tables are falling in scores of countries (including northern China, the US, and India) due to widespread overpumping using powerful diesel and electric pumps. Other countries affected include Pakistan, Afghanistan, and Iran. This will eventually lead to water scarcity and cutbacks in grain harvest. Even with the overpumping of its aquifers, China is developing a grain deficit. When this happens, it will almost certainly drive grain prices upward. Most of the 3 billion people projected to be born worldwide by mid-century will be born in countries already experiencing water shortages. After China and India, there is a second tier of smaller countries with large water deficits – Afghanistan, Algeria, Egypt, Iran, Mexico, and Pakistan. Four of these already import a large share of their grain. Only Pakistan remains self-sufficient. But with a population expanding by 4 million a year, it will likely soon turn to the world market for grain.

Irrigation canals have opened dry desert areas of Egypt to agriculture.

Regionally, Sub-Saharan Africa has the largest number of water-stressed countries of any other place on the globe and as of an estimated 800 million people who live in Africa, 300 million live in a water-stressed environment. It is estimated that by 2030, 75 million to 250 million people in Africa will be living in areas of high water stress, which will likely displace anywhere between 24 million and 700 million people as conditions become increasingly unlivable. Because the majority of Africa remains dependent on an agricultural lifestyle and 80 to 90 percent of all families in rural Africa rely upon producing their own food, water scarcity translates to a loss of food security.

Multimillion-dollar investments beginning in the 1990s by the World Bank have reclaimed desert and turned the Ica Valley in Peru, one of the driest places on earth, into the largest supplier of asparagus in the world. However, the constant irrigation has caused a rapid drop in the water table, in some places as much as eight meters per year, one of the fastest rates of aquifer depletion in the world. The wells of small farmers and local people are beginning to run dry and the water supply for the main city in the valley is under threat. As a cash crop, asparagus has provided jobs for local people, but most of the money goes to the buyers, mainly the British. A 2010 report concluded that the industry is not sustainable and accuses investors, including the World Bank, of failing to take proper responsibility for the effect of their decisions on the water resources of poorer countries. Diverting water from the headwaters of the Ica River to asparagus fields has also led to a water shortage in the mountain region of Huancavelica, where indigenous communities make a marginal living herding.

Land Degradation

Intensive farming often leads to a vicious cycle of exhaustion of soil fertility and decline of agricultural yields. Approximately 40 percent of the world's agricultural land is seriously degraded. In Africa, if current trends of soil degradation continue, the continent might be able to feed just 25 percent of its population by 2025, according to UNU's Ghana-based Institute for Natural Resources in Africa.

Climate Change

Extreme events, such as droughts and floods, are forecast to increase as climate change and global warming takes hold. Ranging from overnight floods to gradually worsening droughts, these will have a range of effects on the agricultural sector. By 2040, almost the entire Nile region, which once included large areas of irrigated agricultural land, is expected to become hot desert where cultivation is impossible due to water limitation. According to the Climate & Development Knowledge Network report *Managing Climate Extremes and Disasters in the Agriculture Sectors: Lessons from the IPCC SREX Report*, the effects will include changing productivity and livelihood patterns, economic losses, and effects on infrastructure, markets and food security. Food security in future will be linked to our ability to adapt agricultural systems to extreme events. For example, the Garifuna women in Honduras are helping to ensure food security locally by reviving and improving production of traditional root crops, building up traditional methods of soil conservation, carrying out training in organic composting and pesticide use and creating the first Garifuna farmers' market. Sixteen towns have worked together to establish tool and seed banks. Efforts to plant wild fruit trees along the coast are helping to prevent soil erosion. The aim is to reduce the communities' vulnerability to the hazards of shifting weather patterns.

Approximately 2.4 billion people live in the drainage basin of the Himalayan rivers. India, China, Pakistan, Afghanistan, Bangladesh, Nepal and Myanmar could experience floods followed by severe droughts in coming decades. In India alone, the Ganges provides water for drinking and farming for more than 500 million people. The west coast of North America, which gets much of its water from glaciers in mountain ranges such as the Rocky Mountains and Sierra Nevada, also would be affected. Glaciers aren't the only worry that the developing nations have; sea level is reported to rise as climate change progresses, reducing the amount of land available for agriculture.

In other parts of the world, a big effect will be low yields of grain according to the World Food Trade Model, specifically in the low latitude regions where much of the developing world is located. From this the price of grain will rise, along with the developing nations trying to grow the grain. Due to this, every 2–2.5% price hike will increase the number of hungry people by 1%. Low crop yields are just one of the problem facing farmers in the low latitudes and tropical regions. The timing and length of the growing seasons, when farmers plant their crops, are going to be changing dramatically, per the USDA, due to unknown changes in soil temperature and moisture conditions.

Another way of thinking about food security and climate change comes from Evan Fraser, a geographer working at the University of Guelph in Ontario Canada. His approach is to explore the vulnerability of food systems to climate change and he defines vulnerability to climate change as situations that occur when relatively minor environmental problems cause major effects on food security. Examples of this include the Irish Potato Famine, which was caused by a rainy year that created ideal conditions for the fungal blight to spread in potato fields, or the Ethiopian Famine in the early 1980s. Three factors stand out as common in such cases, and these three factors act as a diagnostic "tool kit" through which to identify cases where food security may be vulnerable to climate change. These factors are: (1) specialized agro-ecosystems; (2) households with very few livelihood options other than farming; (3) situations where formal institutions do not provide adequate safety nets to protect people. "The International Food Policy Research Institute (IFPRI) estimates that an additional US$ 7.1-7.3 billion per year are needed in agricultural investments to offset the negative effect of climate change on nutrition for children by 2050."

Agricultural Diseases

Diseases affecting livestock or crops can have devastating effects on food availability especially if there are no contingency plans in place. For example, Ug99, a lineage of wheat stem rust which can cause up to 100% crop losses, is present in wheat fields in several countries in Africa and the Middle East and is predicted to spread rapidly through these regions and possibly further afield, potentially causing a wheat production disaster that would affect food security worldwide.

Stripe rust on a wheat stem

The genetic diversity of the crop wild relatives of wheat can be used to improve modern varieties to be more resistant to rust. In their centers of origin wild wheat plants are screened for resistance to rust, then their genetic information is studied and finally wild plants and modern varieties are crossed through means of modern plant breeding in order to transfer the resistance genes from the wild plants to the modern varieties.

Dictatorship and Kleptocracy

Nobel Prize winning economist Amartya Sen has observed that "there is no such thing as an apolitical food problem." While drought and other naturally occurring events may trigger famine conditions, it is government action or inaction that determines its severity, and often even whether or not a famine will occur. The 20th century is full of examples of governments undermining the food security of their own nations – sometimes intentionally.

When governments come to power by force or rigged elections, and not by way of fair and open elections, their base of support is often narrow and built upon cronyism and patronage. Under such conditions "The distribution of food within a country is a political issue. Governments in most countries give priority to urban areas, since that is where the most influential and powerful families and enterprises are usually located. The government often neglects subsistence farmers and rural areas in general. The more remote and underdeveloped the area the less likely the government will be to effectively meet its needs. Many agrarian policies, especially the pricing of agricultural commodities, discriminate against rural areas. Governments often keep prices of basic grains at such artificially low levels that subsistence producers cannot accumulate enough capital to make investments to improve their production. Thus, they are effectively prevented from getting out of their precarious situation."

Further dictators and warlords have used food as a political weapon, rewarding their supporters while denying food supplies to areas that oppose their rule. Under such conditions food becomes a currency with which to buy support and famine becomes an effective weapon to be used against the opposition.

Governments with strong tendencies towards kleptocracy can undermine food security even when harvests are good. When government monopolizes trade, farmers may find that they are free to grow cash crops for export, but under penalty of law only able to sell their crops to government buyers at prices far below the world market price. The government then is free to sell their crop on the world market at full price, pocketing the difference. This creates an artificial "poverty trap" from which even the most hard working and motivated farmers may not escape.

When the rule of law is absent, or private property is non-existent, farmers have little incentive to improve their productivity. If a farm becomes noticeably more productive than neighboring farms, it may become the target of individuals well connected to the government. Rather than risk being noticed and possibly losing their land, farmers may be content with the perceived safety of mediocrity.

As pointed out by William Bernstein in his book *The Birth of Plenty*: "Individuals without property are susceptible to starvation, and it is much easier to bend the fearful and hungry to the will of the state. If a [farmer's] property can be arbitrarily threatened by the state, that power will inevitably be employed to intimidate those with divergent political and religious opinions."

Food Sovereignty

The approach known as food sovereignty views the business practices of multinational corporations as a form of neocolonialism. It contends that multinational corporations have the financial resources available to buy up the agricultural resources of impoverished nations, particularly in the tropics. They also have the political clout to convert these resources to the exclusive production of cash crops for sale to industrialized nations outside of the tropics, and in the process to squeeze the poor off of the more productive lands. Under this view subsistence farmers are left to cultivate only lands that are so marginal in terms of productivity as to be of no interest to the multinational corporations. Likewise, food sovereignty holds it to be true that communities should be able to define their own means of production and that food is a basic human right. With several multinational corporations now pushing agricultural technologies on developing countries, technologies that include improved seeds, chemical fertilizers, and pesticides, crop production has become an increasingly analyzed and debated issue. Many communities calling for food sovereignty are protesting the imposition of Western technologies on to their indigenous systems and agency.

Risks to Food Security

Population Growth

Current UN projections show a continued increase in population in the near future (but a steady decline in the population growth rate), with the global population expected to reach between 8.3 and 10.9 billion by 2050. UN Population Division estimates for the year 2150 range between 3.2 and 24.8 billion; mathematical modeling supports the lower estimate. Some analysts have questioned the sustainability of further world population growth, highlighting the growing pressures on the environment, global food supplies, and energy resources. Solutions for feeding the nine billion in the future are being studied and documented. One out of every seven people on our planet go to sleep hungry. People are suffering due to overpopulation, 25,000 people die of malnutrition and hunger related diseases everyday.

Fossil Fuel Dependence

While agricultural output increased as a result of the Green Revolution, the energy input into the process (that is, the energy that must be expended to produce a crop) has also increased at a greater rate, so that the ratio of crops produced to energy input has decreased over time. Green Revolution techniques also heavily rely on chemical fertilizers, pesticides and herbicides, some of which must be developed from fossil fuels, making agriculture increasingly reliant on petroleum products.

Between 1950 and 1984, as the Green Revolution transformed agriculture around the globe, world grain production increased by 250%. The energy for the Green Revolution was provided by fossil fuels in the form of fertilizers (natural gas), pesticides (oil), and hydrocarbon fueled irrigation.

David Pimentel, professor of ecology and agriculture at Cornell University, and Mario Giampietro, senior researcher at the National Research Institute on Food and Nutrition (INRAN), place in their study *Food, Land, Population and the U.S. Economy* the maximum U.S. population for a sustainable economy at 200 million. To achieve a sustainable economy and avert disaster, the United States must reduce its population by at least one-third, and world population will have to be reduced by two-thirds, says the study.

The authors of this study believe that the mentioned agricultural crisis will only begin to affect us after 2020, and will not become critical until 2050. The oncoming peaking of global oil production (and subsequent decline of production), along with the peak of North American natural gas production will very likely precipitate this agricultural crisis much sooner than expected. Geologist Dale Allen Pfeiffer claims that coming decades could see spiraling food prices without relief and massive starvation on a global level such as never experienced before.

Homogeneity in the Global Food Supply

Since 1961, human diets across the world have become more diverse in the consumption of major commodity staple crops, with a corollary decline in consumption of local or regionally important crops, and thus have become more homogeneous globally. The differences between the foods eaten in different countries were reduced by 68% between 1961 and 2009. The modern "global standard" diet contains an increasingly large percentage of a relatively small number of major staple commodity crops, which have increased substantially in the share of the total food energy (calories), protein, fat, and food weight that they provide to the world's human population, including wheat, rice, sugar, maize, soybean (by +284%), palm oil (by +173%), and sunflower (by +246%). Whereas nations used to consume greater proportions of locally or regionally important crops, wheat has become a staple in over 97% of countries, with the other global staples showing similar dominance worldwide. Other crops have declined sharply over the same period, including rye, yam, sweet potato (by -45%), cassava (by -38%), coconut, sorghum (by -52%) and millets (by -45%). Such crop diversity change in the human diet is associated with mixed effects on food security, improving under-nutrition in some regions but contributing to the diet-related diseases caused by over-consumption of macronutrients.

A small number of major crops, e.g. Soybean, have formed an increasing share of the food energy, protein, fat, and food weight eaten by the world's population over the past 50 years

Price Setting

On April 30, 2008, Thailand, one of the world's biggest rice exporters, announces the project of the creation of the Organisation of Rice Exporting Countries with the potential to develop into a price-fixing cartel for rice. It is a project to organize 21 rice exporting countries to create a homonymous organisation to control the price of rice. The group is mainly made up of Thailand, Vietnam, Cambodia, Laos and Myanmar. The organization attempts to serve the purpose of making a "contribution to ensuring food stability, not just in an individual country but also to address food shortages in the region and the world". However, it is still questionable whether this organization will serve its role as an effective rice price fixing cartel, that is similar to OPEC's mechanism for managing petroleum. Economic analysts and traders said the proposal would go nowhere because of the inability of governments to cooperate with each other and control farmers' output. Moreover, countries that are involved expressed their concern, that this could only worsen the food security.

Land use Change

China needs not less than 120 million hectares of arable land for its food security. China has recently reported a surplus of 15 million hectares. On the other side of the coin, some 4 million hectares of conversion to urban use and 3 million hectares of contaminated land have been reported as well. Furthermore, a survey found that 2.5% of China's arable land is too contaminated to grow food without harm. In Europe, the conversion of agricultural soil implied a net loss of potential. But the rapid loss in the area of arable soils appears to be economically meaningless because EU is perceived to be dependent on internal food supply anymore. During the period 2000-2006 the European Union lost 0.27% of its cropland and 0.26% of its crop productive potential. The loss of agricultural land during the same time was the highest in the Netherlands, which lost 1.57% of its crop production potential within six years. The figures are quite alarming for Cyprus (0.84%), Ireland (0.77%) and Spain (0.49%) as well. In Italy, in the Emilia-Romagna plain (ERP), the conversion of 15,000 ha of agricultural soil (period 2003-2008) implied a net loss of 109,000 Mg per year of wheat, which accounts for the calories needed by 14% of ERP population (425,000 people). Such a loss in wheat production is just 0.02% of gross domestic product (GDP) of the Emilia-Romagna region which is actually a minor effect in financial terms. Additionally, the income from the new land use is often much higher than the one guaranteed by agriculture, as in the case of urbanisation or extraction of raw materials.

Global Catastrophic Risks

As anthropogenic greenhouse gas emissions reduce the stability of the global climate, abrupt climate change could become more intense. The impact of an asteroid or comet larger than about 1 km diameter has the potential to block the sun globally, causing impact winter. Particles in the troposphere would quickly rain out, but particles in the stratosphere, especially sulfate, could remain there for years. Similarly, a supervolcanic eruption would reduce the potential of agricultural production from solar photosynthesis, causing volcanic winter. The Toba super volcanic eruption approximately 70,000 years ago may have nearly caused the extinction of humans (see Toba catastrophe theory). Again, primarily sulfate particles could block the sun for years. Solar blocking is not limited to natural causes as nuclear winter is also possible, which refers to the scenario involving widespread nuclear war and burning of cities that release soot into the stratosphere that would stay there for about 10 years. The high stratospheric temperatures produced by soot absorbing solar radiation would create near-global ozone hole conditions even for a regional nuclear conflict.

Children and Food Security

On April 29, 2008, a UNICEF UK report found that the world's poorest and most vulnerable children are being hit the hardest by climate change. The report, "Our Climate, Our Children, Our Responsibility: The Implications of Climate Change for the World's Children", says that access to clean water and food supplies will become more difficult, particularly in Africa and Asia.

By way of comparison, in one of the largest food producing countries in the world, the United States, approximately one out of six people are "food insecure", including 17 million children, according to the U.S. Department of Agriculture. A 2012 study in the *Journal of Applied Research on Children* found that rates of food security varied significantly by race, class and education. In both kindergarten and third grade, 8% of the children were classified as food insecure, but only 5% of white children were food insecure, while 12% and 15% of black and Hispanic children were food insecure, respectively. In third grade, 13% of black and 11% of Hispanic children are food insecure compared to 5% of white children.

Gender and Food Security

Gender inequality both leads to and is a result of food insecurity. According to estimates women and girls make up 60% of the world's chronically hungry and little progress has been made in ensuring the equal right to food for women enshrined in the Convention on the Elimination of All Forms of Discrimination against Women. Women face discrimination both in education and employment opportunities and within the household, where their bargaining power is lower. On the other hand, gender equality is described as instrumental to ending malnutrition and hunger. Women tend to be responsible for food preparation and childcare within the family and are more likely to spend their income on food and their children's needs. Women also play an important role in food production, processing, distribution and marketing. They often work as unpaid family workers, are involved in subsistence farming and represent about 43% of the agricultural labor force in developing countries, varying from 20% in Latin America to 50% in Eastern and South-eastern Asia and Sub-Saharan Africa. However, women face discrimination in access to land, credit, technologies, finance and other services. Empirical studies suggest that if women had the same access to productive resources as men, women could boost their yields by 20–30%; raising the overall agricultural output in developing countries by 2.5 to 4%. While those are rough estimates, the significant benefit of closing the gender gap on agricultural productivity cannot be denied. The gendered aspects of food security are visible along the four pillars of food security: availability, access, utilization and stability, as defined by the Food and Agriculture Organization.

A Kenyan woman farmer at work in the Mount Kenya region.

The number of people affected by hunger is extremely high, with enormous effects on women and girls. Making this trend disappear "must be a top priority for governments and international institutions". Actions governments take must take into consideration that food insecurity is an issue regarding "equality, rights and social justice". "Food and nutrition insecurity is a political and economic phenomenon fuelled by inequitable global and national processes". Factors like capitalism, exploration of Indigenous lands all contribute to food insecurity for minorities and the people who are the most oppressed in various countries (women being one of these oppressed groups). To emphasis, "food and nutrition insecurity is a gender justice issue". The facts that women and girls are the most oppressed by "the inequitable global economic processes that govern food systems and by global trends such as climate change", shows how institutions continue to place women in positions of disadvantage and impoverishment to make money and thrive on capitalizing the food system. When the government withholds food by raising its prices to amounts only privileged people can afford, they both benefit and are able to control the "lower-class"/ marginalized people via the food market. An interesting fact is that "despite rapid economic growth in India, thousands of women and girls still lack food and nutrition security as a direct result of their lower status compared with men and boys". "Such inequalities are compounded by women and girls' often limited access to productive resources, education and decision-making, by the 'normalised' burden of unpaid work – including care work – and by the endemic problems of gender-based violence (GBV), HIV and AIDS".

Use of Genetically Modified (GM) Crops

One of the most up-and-coming techniques to ensuring global food security is the use of genetically modified (GM) crops. The genome of these crops can be altered to address one or more aspects of the plant that may be preventing it from being grown in various regions under certain conditions. Many of these alterations can address the challenges that were previously mentioned above, including the water crisis, land degradation, and the ever-changing climate.

In agriculture and animal husbandry, the Green Revolution popularized the use of conventional hybridization to increase yield by creating "high-yielding varieties". Often the handful of hybridized breeds originated in developed countries and were further hybridized with local varieties in the rest of the developing world to create high yield strains resistant to local climate and diseases.

The area sown to genetically engineered crops in developing countries is rapidly catching up with the area sown in industrial nations. According to the International Service for the Acquisition of Agri-biotech Applications (ISAAA), GM crops were grown by approximately 8.5 million farmers in 21 countries in 2005; up from 8.25 million farmers in 17 countries in 2004. However, the ISAAA is funded by organisations including prominent agricultural biotechnology corporations, such as Monsanto and Bayer, and there have been several challenges made to the accuracy of ISAAA's global figures.

Opposition to GM Crops

Respected scientists question the safety of biotechnology as a panacea; agroecologists Miguel Altieri and Peter Rosset have enumerated ten reasons why biotechnology will not ensure food security, protect the environment, or reduce poverty. Reasons include:

1. There is no relationship between the prevalence of hunger in a given country and its population
2. Most innovations in agricultural biotechnology have been profit-driven rather than need-driven
3. Ecological theory predicts that the large-scale landscape homogenization with transgenic crops will exacerbate the ecological problems already associated with monoculture agriculture
4. And, that much of the needed food can be produced by small farmers located throughout the world using existing agroecological technologies.

Based on evidence from previous attempts, there is a likely lack of transferability of one type of GM crop from one region to another. For example, modified crops that have proven successful in Asia from the Green Revolution have failed when tried in regions of Africa. More research must be done regarding the specific requirements of growing a specific crop in a specific region.

There is also a drastic lack of education given to governments, farmers, and the community about the science behind GM crops, as well as suitable growing practices. In most relief programs, farmers are given seeds with little explanation and little attention is paid to the resources available to them or even laws that prohibit them from distributing produce. Governments are often not advised on the economic and health implications that come with growing GM crops, and are then left to make judgments on their own. Because they have so little information regarding these crops, they usually shy away from allowing them or do not take the time and effort required to regulate their use. Members of the community that will then consume the produce from these crops are also left in the dark about what these modifications mean and are often scared off by their 'unnatural' origins. This has resulted in failure to properly grow crops as well as strong opposition to the unknown practices.

A study published in June 2016 evaluated the status of the implementation of Golden Rice, which was first developed in the 1990s to produce higher levels of Vitamin A than its non-GMO counterparts. This strain of rice was designed so that malnourished women and children in third world countries who were more susceptible to deficiencies could easily improve their Vitamin A intake levels and prevent blindness, which is a common result. Golden Rice production was centralized to the Philippines, yet there have been many hurdles to jump in order to get production moving. The study showed that the project is far behind schedule and is not living up to its expectations. Although research on Golden Rice still continues, the country has moved forward with other non-GMO initiatives to address the Vitamin A deficiency problem which is so prevasive in that region.

Many anti-GMO activists argue that the use of GM crops decreases biodiversity amongst plants. Livestock biodiversity is also threatened by the modernization of agriculture and the focus on more productive major breeds. Therefore, efforts have been made by governments and non-governmental organizations to conserve livestock biodiversity through strategies such as Cryoconservation of animal genetic resources.

Support of GM Crops

Many GM crop success stories exist, primarily in developed nations like the USA, China, and various countries in Europe. Common GM crops include cotton, maize, and soybeans, all of which are grown throughout North and South America as well as regions of Asia. Modified cotton crops, for example, have been altered such that they are resistant to pests, can grown in more extreme heat, cold, or drought, and produce longer, stronger fibers to be used in textile production.

One of the biggest threats to rice, which is a staple food crop especially in India and other countries within Asia, is blast disease which is a fungal infection that causes lesions to form on all parts of the plant. A genetically engineered strain of rice has been developed so that it is resistant to blast, greatly improving the crop yield of farmers and allowing rice to be more accessible to everyone. Some other crops have been modified such that they produce higher yields per plant or that they require less land for growing. The latter can be helpful in extreme climates with little arable land and also decreases deforestation, as fewer trees need to be cut down in order to make room for crop fields. Others yet have been altered such that they do not require the use of insecticides or fungicides. This addresses various health concerns associated with such pesticides and can also work to improve biodiversity within the area in which these crops are grown.

In a review of Borlaug's 2000 publication entitled *Ending world hunger: the promise of biotechnology and the threat of antiscience zealotry*, the authors argued that Borlaug's warnings were still true in 2010,

GM crops are as natural and safe as today's bread wheat, opined Dr. Borlaug, who also reminded agricultural scientists of their moral obligation to stand up to the antiscience crowd and warn policy makers that global food insecurity will not disappear without this new technology and ignoring this reality global food insecurity would make future solutions all the more difficult to achieve.

—*Rozwadowski and Kagale*

Research conducted by the GMO Risk Assessment and Communication of Evidence (GRACE) program through the EU between 2007 and 2013 focused on many uses of GM crops and evaluated many facets of their effects on human, animal, and environmental health.

Approaches

A liquid manure spreader, equipment that is used to increase agricultural productivity.

By the United Nations

The UN Millennium Development Goals are one of the initiatives aimed at achieving food security in the world. The first Millennium Development Goal states that the UN "is to eradicate extreme hunger and poverty" by 2015. Olivier De Schutter, the UN Special Rapporteur on the Right to Food, advocates for a multidimensional approach to food security challenges. This approach emphasizes the physical availability of food; the social,economic and physical access people have to food; and the nutrition, safety and cultural appropriateness or adequacy of food.

By the Food and Agriculture Organization

The Food and Agriculture Organization of the United Nations stated in *The State of Food Insecurity in the World 2003* that countries that have reduced hunger often had rapid economic growth, specifically in their agricultural sectors. These countries were also characterized as having slower population growth, lower HIV rates, and higher rankings in the Human Development Index. At that time, the FAO considered addressing agriculture and population growth vital to achieving food security. In *The State of Food Insecurity in the World 2012*, the FAO restated its focus on economic growth and agricultural growth to achieve food security and added a focus on the poor and on "nutrition-sensitive" growth. For example, economic growth should be used by governments to provide public services to benefit poor and hungry populations. The FAO also cited smallholders, including women, as groups that should be involved in agricultural growth to generate employment for the poor. For economic and agricultural growth to be "nutrition-sensitive", resources should be utilized to improve access to diverse diets for the poor as well as access to a safe water supply and to healthcare. The FAO has proposed a "twin track" approach to fight food insecurity that combines sustainable development and short-term hunger relief. Development approaches include investing in rural markets and rural infrastructure. In general, the FAO proposes the use of public policies and programs that promote long-term economic growth that will benefit the poor. To obtain short-term food security, vouchers for seeds, fertilizer, or access to services could promote agricultural production. The use of conditional or unconditional food or cash transfers was another approach the FAO noted. Conditional transfers could include school feeding programs, while unconditional transfers could include general food distribution, emergency food aid or cash transfers. A third approach is the use of subsidies as safety nets to increase the purchasing power of households. The FAO stated that "approaches should be human rights-based, target the poor, promote gender equality, enhance long-term resilience and allow sustainable graduation out of poverty."

The FAO noted that some countries have been successful in fighting food insecurity and decreasing the number of people suffering from undernourishment. Bangladesh is an example of a country that has met the Millennium Development Goal hunger target. The FAO credited growth in agricultural productivity and macroeconomic stability for the rapid economic growth in the 1990s that resulted in an increase in food security. Irrigation systems were established through infrastructure development programs. Two programs, HarvestPlus and the Golden Rice Project, provided biofortified crops in order to decrease micronutrient deficiencess.

World Food Day was established on October 16, in honor of the date that the FAO was founded in 1945. On this day, the FAO hosts a variety of event at the headquarters in Rome and around the world, as well as seminars with UN officials.

By the World Food Programme

The World Food Programme (WFP) is an agency of the United Nations that uses food aid to promote food security and eradicate hunger and poverty. In particular, the WFP provides food aid to refugees and to others experiencing food emergencies. It also seeks to improve nutrition and quality of life to the most vulnerable populations and promote self-reliance. An example of a WFP program is the "Food For Assets" program in which participants work on new infrastructure, or learn new skills, that will increase food security, in exchange for food. The WFP and the Government of Kenya have partnered in the Food For Assets program in hopes of increasing the resilience of communities to shocks.

Fight Hunger: Walk the World campaign is a United Nations World Food Programme initiative.

Global Partnerships to Achieve Food Security and End Hunger

In April 2012, the Food Assistance Convention was signed, the world's first legally binding international agreement on food aid. The May 2012 Copenhagen Consensus recommended that efforts to combat hunger and malnutrition should be the first priority for politicians and private sector philanthropists looking to maximize the effectiveness of aid spending. They put this ahead of other priorities, like the fight against malaria and AIDS.

The main global policy to reduce hunger and poverty are the recently approved Sustainable Development Goals. In particular Goal 2: Zero Hunger sets globally agreed targets to end hunger, achieve food security and improved nutrition and promote sustainable agriculture by 2030. A number of organizations have formed initiatives with the more ambitious goal to achieve this outcome in only 10 years, by 2025:

- In 2013 Caritas International started a Caritas-wide initiative aimed at ending systemic hunger by 2025. The One human family, food for all campaign focuses on awareness raising, improving the effect of Caritas programs and advocating the implementation of the Right to Food.
- The partnership Compact2025, led by IFPRI with the involvement of UN organisations, NGOs and private foundations develops and disseminates evidence-based advice to politicians and other decision-makers aimed at ending hunger and undernutrition in the coming 10 years, by 2025. It bases its claim that hunger can be ended by 2025 on a report by Shenggen Fan and Paul Polman that analyzed the experiences from China, Vietnam, Brazil and Thailand and concludes that eliminating hunger and undernutrition was possible by 2025.
- In June 2015, the European Union and the Bill & Melinda Gates Foundation have launched a partnership to combat undernutrition especially in children. The program will initiatilly be implemented in Bangladesh, Burundi, Ethiopia, Kenya, Laos and Niger and will help these countries to improve information and analysis about nutrition so they can develop effective national nutrition policies.
- The Food and Agriculture Organization of the UN has created a partnership that will act through the African Union's CAADP framework aiming to end hunger in Africa by 2025. It includes different interventions including support for improved food production, a strengthening of social protection and integration of the Right to Food into national legislation.

By the United States Agency for International Development

The United States Agency for International Development (USAID) proposes several key steps to increasing agricultural productivity which is in turn key to increasing rural income and reducing food insecurity. They include:

- Boosting agricultural science and technology. Current agricultural yields are insufficient to feed the growing populations. Eventually, the rising agricultural productivity drives economic growth.
- Securing property rights and access to finance.
- Enhancing human capital through education and improved health.
- Conflict prevention and resolution mechanisms and democracy and governance based on principles of accountability and transparency in public institutions and the rule of law are basic to reducing vulnerable members of society.

Since the 1960s, the U.S. has been implementing a Food Stamp Program (now called the Supplemental Nutrition Assistance Program) to directly target consumers who lack the income to purchase food. According to Tim Josling, a Senior Fellow at the Freeman Spogli Institute for International Studies, Stanford University, food stamps or other methods of distribution of purchasing power directly to consumers might fit into the range of international programs under consideration to tackle food insecurity.

Improving Agricultural Productivity to Benefit the Rural Poor

There are strong, direct relationships between agricultural productivity, hunger, poverty, and sustainability. Three-quarters of the world's poor live in rural areas and make their living from agriculture. Hunger and child malnutrition are greater in these areas than in urban areas. Moreover, the higher the proportion of the rural population that obtains its income solely from subsistence farming (without the benefit of pro-poor technologies and access to markets), the higher the incidence of malnutrition. Therefore, improvements in agricultural productivity aimed at small-scale farmers will benefit the rural poor first. Food and feed crop demand is likely to double in the next 50 years, as the global population approaches nine billion. Growing sufficient food will require people to make changes such as increasing productivity in areas dependent on rainfed agriculture; improving soil fertility management; expanding cropped areas; investing in irrigation; conducting agricultural trade between countries; and reducing gross food demand by influencing diets and reducing post-harvest losses.

According to the Comprehensive Assessment of Water Management in Agriculture, a major study led by the International Water Management Institute (IWMI), managing rainwater and soil moisture more effectively, and using supplemental and small-scale irrigation, hold the key to helping the greatest number of poor people. It has called for a new era of water investments and policies for upgrading rainfed agriculture that would go beyond controlling field-level soil and water to bring new freshwater sources through better local management of rainfall and runoff. Increased agricultural productivity enables farmers to grow more food, which translates into better diets and, under market conditions that offer a level playing field, into higher farm incomes. With more money, farmers are more likely to diversify production and grow higher-value crops, benefiting not only themselves but the economy as a whole."

A farmer on the outskirts of Lilongwe (Malawi) prepares a field for planting.

Researchers suggest forming an alliance between the emergency food program and community-supported agriculture, as some countries' food stamps cannot be used at farmer's markets and places where food is less processed and grown locally. The gathering of wild food plants appears to be an efficient alternative method of subsistence in tropical countries, which may play a role in poverty alleviation.

Producing Food without Agriculture

David Denkenberger and Joshua Pearce have proposed in Feeding Everyone No Matter What a variety of alternate foods which convert fossil fuels or biomass into food without sunlight to address sunlight-blocking food security scenarios. The solution using fossil fuel energy source is natural gas digesting bacteria. One example of a biomass alternate food is that mushrooms can grow directly on wood without sunlight. Another example is that cellulosic biofuel production typically already creates sugar as an intermediate product.

Large-scale Food Stockpiling

The minimum annual global wheat storage is approximately 2 months. To counteract the severe food security issues caused by global catastrophic risks, years of food storage has been proposed. Though this could ameliorate smaller scale problems like regional conflict and drought, it would exacerbate current food insecurity by raising food prices.

Agricultural Insurances

Insurance is a financial instrument, which allows exposed individuals to pool resources to spread their risk. They do so by contributing premium to an insurance fund, which will indemnify those who suffer insured loss. This procedure reduces the risk for an individual by spreading his/her risk among the multiple fund contributors. Insurance can be designed to protect many types of individuals and assets against single or multiple perils and buffer insured parties against sudden and dramatic income or asset loss.

Crop insurance is purchased by agricultural producers to protect themselves against either the loss of their crops due to natural disasters. Two type of insurances are available: (1) claim-based insurances, and (2) index-based insurances. In particular in poor countries facing food security problems, index-based insurances offer some interesting advantages: 1) indices can be derived from globally available satellite images that correlate well with what is insured; (2) these indices can be delivered at low cost; and (3) the insurance products open up new markets that are not served by claim-based insurances.

An advantage of index-based insurance is that it can potentially be delivered at lower cost. A significant barrier that hinders uptake of claim-based insurance is the high transaction cost for searching for prospective policyholders, negotiating and administering contracts, verifying losses and determining payouts. Index insurance eliminates the loss verification step, thereby mitigating a significant transaction cost. A second advantage of index-based insurance is that, because it pays an indemnity based on the reading of an index rather than individual losses, it eliminates much of the fraud, moral hazard and adverse selection, which are common in classical claim-based insurance. A further advantage of index insurance is that payments based on a standardized and indisputable index also allow for a fast indemnity payment. The indemnity payment could be automated, further reducing transaction costs.

Basis risk is a major disadvantage of index-based insurance. It is the situation where an individual experiences a loss without receiving payment or vice versa. Basis risk is a direct result of the strength of the relation between the index that estimates the average loss by the insured group and the loss of insured assets by an individual. The weaker this relation the higher the basis risk. It is obvious that high basis risk undermines the willingness of potential clients to purchase insurance. It thus challenges insurance companies to design insurances such as to minimize basis risk.

Food Security, Quality through Bees

Bees and other pollinating insects are currently improving the food production of 2 billion small farmers worldwide, helping to ensure food security for the world's population. Research shows that if pollination is managed well on small diverse farms, with all other factors being equal, crop yields can increase by a significant median of 24 percent.

Criticism

As of 2015 the concept of food security has mostly focused on food calories rather than the quality and nutrition of food. The concept of nutrition security evolved over time. In 1995, it has been defined as "adequate nutritional status in terms of protein, energy, vitamins, and minerals for all household members at all times".

How animal pollinators positively affect fruit condition and nutrient content is still being discovered.

Food Safety

Food safety is a scientific discipline describing handling, preparation, and storage of food in ways that prevent foodborne illness. This includes a number of routines that should be followed to avoid potentially severe health hazards. In this way food safety often overlaps with food defense to prevent harm to consumers. The tracks within this line of thought are safety between industry and the market and then between the market and the consumer. In considering industry to market practices, food safety considerations include the origins of food including the practices relating to food labeling, food hygiene, food additives and pesticide residues, as well as policies on biotechnology and food and guidelines for the management of governmental import and export inspection and certification systems for foods. In considering market to consumer practices, the usual thought is that food ought to be safe in the market and the concern is safe delivery and preparation of the food for the consumer.

Food can transmit disease from person to person as well as serve as a growth medium for bacteria that can cause food poisoning. In developed countries there are intricate standards for food preparation, whereas in lesser developed countries the main issue is simply the availability of adequate safe water, which is usually a critical item. In theory, food poisoning is 100% preventable. The five key principles of food hygiene, according to WHO, are:

1. Prevent contaminating food with pathogens spreading from people, pets, and pests.
2. Separate raw and cooked foods to prevent contaminating the cooked foods.
3. Cook foods for the appropriate length of time and at the appropriate temperature to kill pathogens.
4. Store food at the proper temperature.
5. Do use safe water and safe raw materials.

Issues

Food safety issues and regulations concern:

- Agriculture and animal husbandry practices

- Food manufacturing practices
- Food additives
- Novel foods
- Genetically modified foods
- Food labels

ISO 22000

ISO 22000 is a standard developed by the International Organization for Standardization dealing with food safety. This is a general derivative of ISO 9000. ISO 22000 standard: The ISO 22000 international standard specifies the requirements for a food safety management system that involves interactive communication, system management, prerequisite programs, HACCP principles.

Incidence

A 2003 World Health Organization (WHO) report concluded that about 30% of reported food poisoning outbreaks in the WHO European Region occur in private homes. According to the WHO and CDC, in the USA alone, annually, there are 76 million cases of foodborne illness leading to 325,000 hospitalizations and 5,000 deaths.

Regulations by Jurisdiction and Agency

WHO and FAO

In 2003, the WHO and FAO published the Codex Alimentarius which serves as an guideline to food safety.

However, according to Unit 04 - Communication of Health & Consumers Directorate-General of the European Commission (SANCO): "The Codex, while being recommendations for voluntary application by members, Codex standards serve in many cases as a basis for national legislation. The reference made to Codex food safety standards in the World Trade Organizations' Agreement on Sanitary and Phytosanitary measures (SPS Agreement) means that Codex has far reaching implications for resolving trade disputes. WTO members that wish to apply stricter food safety measures than those set by Codex may be required to justify these measures scientifically." So, an agreement made in 2003, signed by all member states, inclusive all EU, in the codex Stan Codex 240 – 2003 for coconut milk, sulphite containing additives like E223 and E 224 are allowed till 30 mg/kg, does NOT mean, they are allowed into the EU, see RASFF entries from Denmark: 2012.0834; 2011.1848; en 2011.168, "sulphite unauthorised in coconut milk from Thailand ". Same for polysorbate E 435: see 2012.0838 from Denmark, unauthorised polysorbates in coconut milk and, 2007.AIC from France. Only for the latter the EU amended its regulations with (EU) No 583/2012 per 2 July 2012 to allow this additive, already used for decades and absolutely necessary.

Australia

Food Standards Australia New Zealand requires all food businesses to implement food safety systems. These systems are designed to ensure food is safe to consume and halt the increasing incidence of food poisoning, and they include basic food safety training for at least one person in each business. Food safety training is delivered in various forms by, among other organisations, Registered Training Organizations (RTOs), after which staff are issued a nationally recognised unit of competency code on their certificate. Basic food safety training includes:

- Understanding the hazards associated with the main types of food and the conditions to prevent the growth of bacteria which can cause food poisoning and to prevent illness.
- Potential problems associated with product packaging such as leaks in vacuum packs, damage to packaging or pest infestation, as well as problems and diseases spread by pests.
- Safe food handling. This includes safe procedures for each process such as receiving, re-packing, food storage, preparation and cooking, cooling and re-heating, displaying products, handling products when serving customers, packaging, cleaning and sanitizing, pest control, transport and delivery. Also covers potential causes of cross contamination.
- Catering for customers who are particularly at risk of food-borne illness, as well as those with allergies or intolerance.
- Correct cleaning and sanitizing procedures, cleaning products and their correct use, and the storage of cleaning items such as brushes, mops and cloths.
- Personal hygiene, hand washing, illness, and protective clothing.

Food safety standards and requirements are set out at the national level in the Food Standards Code, and brought Into force in each state by state-based Acts and Regulations. Legislation means that people responsible for selling or serving unsafe food may be liable for heavy fines.

China

Food safety is a growing concern in Chinese agriculture. The Chinese government oversees agricultural production as well as the manufacture of food packaging, containers, chemical additives, drug production, and business regulation. In recent years, the Chinese government attempted to consolidate food regulation with the creation of the State Food and Drug Administration in 2003, and officials have also been under increasing public and international pressure to solve food safety problems. However, it appears that regulations are not well known by the trade. Labels used for "green" food, "organic" food and "pollution-free" food are not well recognized by traders and many are unclear about their meaning. A survey by the World Bank found that supermarket managers had difficulty in obtaining produce that met safety requirements and found that a high percentage of produce did not comply with established standards.

Traditional marketing systems, whether in China or the rest of Asia, presently provide little motivation or incentive for individual farmers to make improvements to either quality or safety as their produce tends to get grouped together with standard products as it progresses through the marketing channel. Direct linkages between farmer groups and traders or ultimate buyers, such as supermarkets, can help avoid this problem. Governments need to improve the condition of many markets through upgrading management and reinvesting market fees in physical infrastructure. Wholesale markets need to investigate the feasibility of developing separate sections to handle fruits and vegetables that meet defined safety and quality standards.

European Union

The parliament of the European Union (EU) makes legislation in the form of directives and regulations, many of which are mandatory for member states and which therefore must be incorporated into individual countries' national legislation. As a very large organisation that exists to remove barriers to trade between member states, and into which individual member states have only a proportional influence, the outcome is often seen as an excessively bureaucratic 'one size fits all' approach. However, in relation to food safety the tendency to err on the side of maximum protection for the consumer may be seen as a positive benefit. The EU parliament is informed on food safety matters by the European Food Safety Authority.

Individual member states may also have other legislation and controls in respect of food safety, provided that they do not prevent trade with other states, and can differ considerably in their internal structures and approaches to the regulatory control of food safety.

From 13 December 2014, new legislation - the EU Food Information for Consumers Regulation 1169/2011 - require food businesses to provide allergy information on food sold unpackaged, in for example catering outlets, deli counters, bakeries and sandwich bars.

France

Agence nationale de sécurité sanitaire de l'alimentation, de l'environnement et du travail (anses) is a French governmental agency dealing with food safety.

Germany

The Federal Ministry of Food, Agriculture and Consumer Protection (BMEL) is a Federal Ministry of the Federal Republic of Germany. History: Founded as Federal Ministry of Food, Agriculture and Foresting in 1949, this name did not change until 2001. Then the name changed to Federal Ministry of Consumer Protection, Food and Agriculture. At the 22nd of November 2005, the name got changed again to its current state: Federal Ministry of Food, Agriculture and Consumer Protection. The reason for this last change was that all the resorts should get equal ranking which was achieved by sorting the resorts alphabetically. Vision: A balanced and healthy diet with safe food, distinct consumer rights and consumer information for various areas of life, and a strong and sustainable agriculture as well as perspectives for our rural areas are important goals of the Federal Ministry of Food, Agriculture and Consumer Protection (BMELV). The Federal Office of Consumer Protection and Food Safety is under the control of the Federal Ministry of Food, Agriculture and Consumer Protection. It exercises several duties, with which it contributes to safer food and thereby intensifies health-based consumer protection in Germany. Food can be manufactured and sold within Germany without a special permission, as long as it does not cause any damage on consumers' health and meets the general standards set by the legislation. However, manufacturers, carriers, importers and retailers are responsible for the food they pass into circulation. They are obliged to ensure and document the safety and quality of their food with the use of in-house control mechanisms.

Hong Kong

In Hong Kong SAR, the Centre for Food Safety is in charge of ensuring food sold is safe and fit for consumption.

India

Food Safety and Standards Authority of India, established under the Food Safety and Standards Act, 2006, is the regulating body related to food safety and laying down of standards of food in India.

New Zealand

The New Zealand Food Safety Authority (NZFSA), or Te Pou Oranga Kai O Aotearoa is the New Zealand government body responsible for food safety. NZFSA is also the controlling authority for imports and exports of food and food-related products. The NZFSA as of 2012 is now a division of the Ministry for Primary Industries (MPI) and is no longer its own organization.

Pakistan

Pakistan does not have an integrated legal framework but has a set of laws, which deals with various aspects of food safety. These laws, despite the fact that they were enacted long time ago, have tremendous capacity to achieve at least minimum level of food safety. However, like many other laws, these laws remain very poorly enforced. There are four laws that specifically deal with food safety. Three of these laws directly focus issues related to food safety, while the fourth, the Pakistan Standards and Quality Control Authority Act, is indirectly relevant to food safety.

The Pure Food Ordinance 1960 consolidates and amends the law in relation to the preparation and the sale of foods. All provinces and some northern areas have adopted this law with certain amendments. Its aim is to ensure purity of food being supplied to people in the market and, therefore, provides for preventing adulteration. The Pure Food Ordinance 1960 does not apply to cantonment areas. There is a separate law for cantonments called "The Cantonment Pure Food Act, 1966". There is no substantial difference between the Pure Food Ordinance 1960 and The Cantonment Pure Food Act. Even the rules of operation are very much similar.

Pakistan Hotels and Restaurant Act, 1976 applies to all hotels and restaurants in Pakistan and seeks to control and regulate the rates and standard of service(s) by hotels and restaurants. In addition to other provisions, under section 22(2), the sale of food or beverages that are contaminated, not prepared hygienically or served in utensils that are not hygienic or clean is an offense. There are no express provisions for consumer complaints in the Pakistan Restaurants Act, 1976, Pakistan Penal Code, 1860 and Pakistan Standards and Quality Control Authority Act, 1996. The laws do not prevent citizens from lodging complaints with the concerned government officials; however, the consideration and handling of complaints is a matter of discretion of the officials.

South Korea

Korea Food & Drug Administration

Korea Food & Drug Administration (KFDA) is working for food safety since 1945. It is part of the Government of South Korea.

IOAS-Organic Certification Bodies Registered in KFDA: "Organic" or related claims can be labelled on food products when organic certificates are considered as valid by KFDA. KFDA admits organic certificates which can be issued by 1) IFOAM (International Federation of Organic Agriculture Movement) accredited certification bodies 2) Government accredited certification bodies – 328 bodies in 29 countries have been registered in KFDA.

Food Import Report: According to Food Import Report, it is supposed to report or register what you import. Competent authority is as follows:

Product	Authority
Imported Agricultural Products, Processed Foods, Food Additives, Utensils, Containers & Packages or Health Functional Foods	KFDA (Korea Food and Drug Administration)
Imported Livestock, Livestock products (including Dairy products)	NVRQS (National Veterinary Research and Quarantine Service)
Packaged meat, milk & dairy products (butter, cheese), hamburger patties, meat ball and other processed products which are stipulated by Livestock Sanitation Management Act	NVRQS (National Veterinary Research and Quarantine Service)
Imported Marine products; fresh, chilled, frozen, salted, dehydrated, eviscerated marine produce which can be recognized its characteristics	NFIS (National Fisheries Products Quality Inspection Service)

National Institute of Food and Drug Safety Evaluation

National Institute of Food and Drug Safety Evaluation (NIFDS) is functioning as well. The National Institute of Food and Drug Safety Evaluation is a national organization for toxicological tests and research. Under the Korea Food & Drug Administration, the Institute performs research on toxicology, pharmacology, and risk analysis of foods, drugs, and their additives. The Institute strives primarily to understand important biological triggering mechanisms and improve assessment methods of human exposure, sensitivities, and risk by (1) conducting basic, applied, and policy research that closely examines biologically triggering harmful effects on the regulated products such as foods, food additives, and drugs, and operating the national toxicology program for the toxicological test development and inspection of hazardous chemical substances assessments. The Institute ensures safety by investigation and research on safety by its own researchers, contract research by external academicians and research centers.

Taiwan

In Taiwan, the Ministry of Health and Welfare in charge of Food and Drug Safety, also evaluate the catering industry to maintenance the food product quality. Currently, US$29.01 million of budget is allocated each year for food safety-related efforts.

United Kingdom

In the UK the Food Standards Agency is an independent government department responsible for food safety and hygiene across the UK. They work with businesses to help them produce safe food, and with local authorities to enforce food safety regulations. In 2006 food hygiene legislation changed and new requirements came into force. The main requirement resulting from this change is that if you own or run a food business in the UK, you must have a documented Food Safety Management System, which is based on the principles of Hazard Analysis Critical Control Point HACCP.

United States

The US food system is regulated by numerous federal, state and local officials. It has been criticized as lacking in "organization, regulatory tools, and not addressing food borne illness."

Federal Level Regulation

The Food and Drug Administration publishes the Food Code, a model set of guidelines and procedures that assists food control jurisdictions by providing a scientifically sound technical and legal basis for regulating the retail and food service industries, including restaurants, grocery stores and institutional foodservice providers such as nursing homes. Regulatory agencies at all levels of government in the United States use the FDA Food Code to develop or update food safety rules in their jurisdictions that are consistent with national food regulatory policy. According to the FDA, 48 of 56 states and territories, representing 79% of the U.S. population, have adopted food codes patterned after one of the five versions of the Food Code, beginning with the 1993 edition.

In the United States, federal regulations governing food safety are fragmented and complicated, according to a February 2007 report from the Government Accountability Office. There are 15 agencies sharing oversight responsibilities in the food safety system, although the two primary agencies are the U.S. Department of Agriculture (USDA) Food Safety and Inspection Service (FSIS), which is responsible for the safety of meat, poultry, and processed egg products, and the Food and Drug Administration (FDA), which is responsible for virtually all other foods.

The Food Safety and Inspection Service has approximately 7,800 inspection program personnel working in nearly 6,200 federally inspected meat, poultry and processed egg establishments. FSIS is charged with administering and enforcing the Federal Meat Inspection Act, the Poultry Products Inspection Act, the Egg Products Inspection Act, portions of the Agricultural Marketing Act, the Humane Slaughter Act, and the regulations that implement these laws. FSIS inspection program personnel inspect every animal before slaughter, and each carcass after slaughter to ensure public health requirements are met. In fiscal year (FY) 2008, this included about 50 billion pounds of livestock carcasses, about 59 billion pounds of poultry carcasses, and about 4.3 billion pounds of processed egg products. At U.S. borders, they also inspected 3.3 billion pounds of imported meat and poultry products.

Industry Pressure

There have been concerns over the efficacy of safety practices and food industry pressure on U.S. regulators. A study reported by Reuters found that "the food industry is jeopardizing U.S. public health by withholding information from food safety investigators or pressuring regulators to withdraw or alter policy designed to protect consumers". A survey found that 25% of U.S. government inspectors and scientists surveyed have experienced during the past year corporate interests forcing their food safety agency to withdraw or to modify agency policy or action that protects consumers. Scientists have observed that management undercuts field inspectors who stand up for food safety against industry pressure. According to Dr. Dean Wyatt, a USDA veterinarian who oversees federal slaughter house inspectors, "Upper level management does not adequately support field inspectors and the actions they take to protect the food supply. Not only is there lack of support, but there's outright obstruction, retaliation and abuse of power." A growing number of food and beverage manufacturers are improving food safety standards by incorporating a food safety management system which automates all steps in the food quality management process.

State and Local Regulation

A number of U.S. states have their own meat inspection programs that substitute for USDA inspection for meats that are sold only in-state. Certain state programs have been criticized for undue leniency to bad practices.

However, other state food safety programs supplement, rather than replace, Federal inspections, generally with the goal of increasing consumer confidence in the state's produce. For example, state health departments have a role in investigating outbreaks of food-borne disease bacteria, as in the case of the 2006 outbreak of *Escherichia coli* O157:H7 (a pathogenic strain of the ordinarily harmless bacteria, *E. coli*) from processed spinach. Health departments also promote better food processing practices to eliminate these threats.

In addition to the US Food and Drug Administration, several states that are major producers of fresh fruits and vegetables (including California, Arizona and Florida) have their own state programs to test produce for pesticide residues.

Restaurants and other retail food establishments fall under state law and are regulated by state or local health departments. Typically these regulations require official inspections of specific design features, best food-handling practices, and certification of food handlers. In some places a letter grade or numerical score must be prominently posted following each inspection. In some localities, inspection deficiencies and remedial action are posted on the Internet.

Vietnam

The Vietnam Food Administration is responsible for managing food hygiene, safety, and quality and has made significant progress since its establishment in 1999. Food safety remains a high priority in Vietnam with the growth of export markets and increasing food imports raising the need to rapidly build capacity of the Food Administration in order to reduce threats of foodborne disease. The Food Administration has demonstrated commitment to the food safety challenges it faces, and has embarked on an innovative capacity building activity with technical assistance from the World Health Organization.

Manufacturing Control

HACCP Guidelines

United Kingdom

Foodstuffs in the UK have one of two labels to indicate the nature of the deterioration of the product and any subsequent health issues. EHO Food Hygiene certification is required to prepare and distribute food. While there is no specified expiry date of such a qualification the changes in legislation it is suggested to update every five years.

Best before indicates a future date beyond which the food product *may* lose quality in terms of taste or texture amongst others, but does not imply any serious health problems if food is consumed beyond this date (within reasonable limits).

Use by indicates a legal date beyond which it is not permissible to sell a food product (usually one that deteriorates fairly rapidly after production) due to the potential serious nature of consumption of pathogens. Leeway is sometimes provided by producers in stating display until dates so that products are not at their limit of safe consumption on the actual date stated (this latter is voluntary and not subject to regulatory control). This allows for the variability in production, storage and display methods.

United States

With the exception of infant formula and baby foods which must be withdrawn by their expiration date, Federal law does not require expiration dates. For all other foods, except dairy products in some states, freshness dating is strictly voluntary on the part of manufacturers. In response to consumer demand, perishable foods are typically labelled with a Sell by date. It is up to the consumer to decide how long after the Sell by date a package is usable. Other common dating statements are Best if used by, Use-by date, Expiration date, Guaranteed fresh <date>, and Pack date.

Australia and New Zealand

Guide to Food Labelling and Other Information Requirements: This guide provides background information on the general labelling requirements in the Code. The information in this guide applies both to food for retail sale and to food for catering purposes. Foods for catering purposes means those foods for use in restaurants, canteens, schools, caterers or self-catering institutions, where food is offered for immediate consumption. Labelling and information requirements in the new Code apply both to food sold or prepared for sale in Australia and New Zealand and food imported into Australia and New Zealand. Warning and Advisory Declarations, Ingredient Labelling, Date Marking, Nutrition Information Requirements, Legibility Requirements for Food Labels, Percentage Labelling, Information Requirements for Foods Exempt from Bearing a Label.

Food Grading

Food grading involves the inspection, assessment and sorting of various foods regarding quality, freshness, legal conformity and market value. Food grading often occurs by hand, in which foods are assessed and sorted. Machinery is also used to grade foods, and may involve sorting products by size, shape and quality. For example, machinery can be used to remove spoiled food from fresh product.

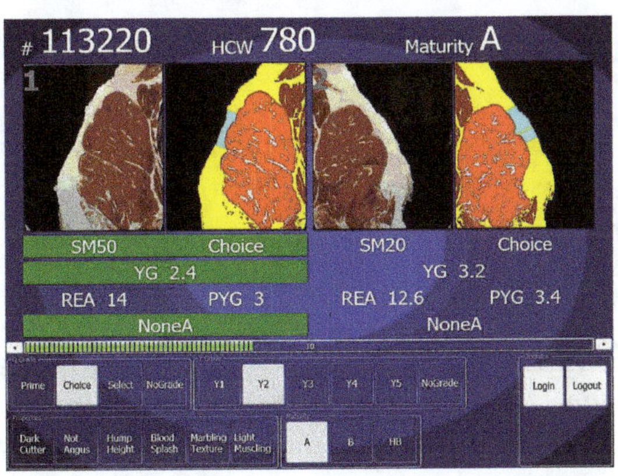

A screenshot from the electronic grading system showing USDA Choice, Yield Grade 2 beef. The left is the natural color view of the cut; the right is the instrument enhanced view that details the amount of marbling, size, and fat thickness.

Optical sorting achieves non-destructive, 100 percent inspection in-line at full production volumes

By Food Type

Beef

Beef grading in the United States is performed by the United States Department of Agriculture's (USDA) Agricultural and Marketing Service. There are eight beef quality grades, with U.S. Prime being the highest grade and U.S. Canner being the lowest grade. Beef grading is a complex and detailed process, requiring graders to think and calculate quickly with great accuracy. Using technology to complement and supplement the onsite human graders generates an efficient and more precise process.

Inspected beef carcasses tagged by the USDA

Beer

In beer grading, the letter "X" is used on some beers, and was traditionally a mark of beer strength, with the more Xs the greater the strength. Some sources suggest that the origin of the mark was in the breweries of medieval monasteries Another plausible explanation is contained in a treatise entitled "The Art of Brewing" published in London in 1829. It says; *"The duties on ale and beer, which were first imposed in 1643... at a certain period, in distinguishing between small beer and strong, all ale or beer, sold at or above ten shillings per barrel, was reckoned to be* strong *and was, therefore, subjected to a higher duty. The cask which contained this strong beer was then first marked with an X signifying ten; and hence the present quack-like denominations of XX (double X) and XXX (treble X) on the casks and accounts of the strong-ale brewers".*

In mid-19th century England, the use of "X" and other letters had evolved into a standardised grading system for the strength of beer. Today, it is used as a trade mark by a number of brewers in the United Kingdom, the Commonwealth and the United States.

European Bitterness Units scale, often abbreviated as EBU, is a scale for measuring the perceived bitterness of beer, with lower values being generally "less bitter" and higher values "more bitter". The scale and method are defined by the European Brewery Convention, and the numerical value should be the same as of the International Bitterness Units scale (IBU), defined in co-operation with the American Society of Brewing Chemists. However, the exact process of determining EBU and IBU values differs slightly, which may in theory result with slightly smaller values for EBU than IBU.

The International Bittering Units scale, or simply IBU scale, provides a measure of the bitterness of beer, which is provided by the hops used during brewing. Bittering units are measured through the use of a spectrophotometer and solvent extraction.

Butter

Coconut Milk

Several grades of coconut milk exist: from thick at 20-22% fat to thin at 5-7% fat level.

Coffee

Coffee growers, traders, and roasters grade beans based on size, color, and a variety of other characteristics. Coffees of exceptional quality are traded as "specialty coffees" and fetch a higher price in the international market.

Dark-roasted coffee beans

After the roast, Coffee grading involves assessment of roasted coffee seed colorization and then labeling as light, medium light, medium, medium dark, dark, or very dark. A more accurate method of discerning the degree of roast involves measuring the reflected light from roasted seeds illuminated with a light source in the near infrared spectrum. This elaborate light meter uses a process known as spectroscopy to return a number that consistently indicates the roasted coffee's relative degree of roast or flavor development.

Eggs

In the United States, egg grading is performed by the USDA, and is based upon the interior quality of the egg (see Haugh unit) and the appearance and condition of the egg shell. Eggs of any quality grade may differ in weight (size). Egg grading is performed by candling, which involves observing the interior of eggs by placing them in front of a bright light.

Guar Gum

Guar gum grading involves analysis for coloration, viscosity, moisture, granulation, protein content and insolubles ash.

Honey

Honey grading in the United States is performed voluntarily based upon USDA standards (USDA does offer inspection and grading "as on-line (in-plant) or lot inspection...upon application, on a fee-for-service basis.") . Honey is graded based upon a number of factors, including water content, flavor and aroma, absence of defects and clarity. Honey is also classified by color though it is not a factor in the grading scale. U.S. honey grade scales are Grade A, Grade B, Grade C and Grade substandard.

Lobster

A lobster

In the U.S., lobster grading involves denoting lobsters as new-shell, hard-shell or old-shell, and because lobsters which have recently shed their shells are the most delicate, there is an inverse relationship between the price of American lobster and its flavour. New-shell lobsters have paper-thin shells and a worse meat-to-shell ratio, but the meat is very sweet. However, the lobsters are so delicate that even transport to Boston almost kills them, making the market for new-shell lobsters strictly local to the fishing towns where they are offloaded. Hard-shell lobsters with firm shells, but with less sweet meat, can survive shipping to Boston, New York and even Los Angeles, so they command a higher price than new-shell lobsters. Meanwhile, old-shell lobsters, which have not shed since the previous season and have a coarser flavour, can be air-shipped anywhere in the world and arrive alive, making them the most expensive. One seafood guide notes that an eight-dollar lobster dinner at a restaurant overlooking fishing piers in Maine is consistently delicious, while "the eighty-dollar lobster in a three-star Paris restaurant is apt to be as much about presentation as flavor".

Maple Syrup

In Canada, the grading of maple syrup involves classification by the Canadian Food Inspection Agency (CFIA) of maple syrups into one of three grades, each with several colour classes: Canada No. 1, including Extra Light, Light, and Medium; No. 2 Amber; and finally No. 3 Dark or any other ungraded category. Producers in Quebec and Ontario may follow either federal or provincial grading guidelines, which differ slightly.

Maple syrup from Quebec, graded as No. 1, Light (enlarge image to read the label atop the can)

The United States uses different grading standards. Maple syrup is divided into two major grades: Grade A and Grade B. Grade A is further divided into three subgrades: Light Amber (sometimes known as Fancy), Medium Amber, and Dark Amber. The Vermont Agency of Agriculture Food and Markets uses a similar grading system of color, and is roughly equivalent, especially for lighter syrups, but using letters: "AA", "A", etc. The Vermont grading system differs from the US system in maintaining a slightly higher standard of product density (measured on the Baumé scale). New Hampshire maintains a similar standard, but not a separate state grading scale. The Vermont-graded product has 0.9 percent more sugar and less water in its composition than US-graded. One grade of syrup not for table use, called commercial or Grade C, is also produced under the Vermont system. Vermont inspectors enforce strict syrup grading regulations, and can fine producers up to US$1000 for labelling syrup incorrectly.

Milk

In the United States, there are two grades of milk, with Grade A primarily used for direct sales and consumption in stores, and Grade B used for indirect consumption, such as in cheese making or other processing.

The differences between the two grades are defined in the Wisconsin administrative code for Agriculture, Trade, and Consumer Protection, chapter 60. Grade B generally refers to milk that is cooled in milk cans, which are immersed in a bath of cold flowing water that typically is drawn up from an underground water well rather than using mechanical refrigeration.

Oranges

The USDA has established the following grades for Florida oranges, which primarily apply to oranges sold as fresh fruit: US Fancy, US No. 1 Bright, US No. 1, US No. 1 Golden, US No. 1 Bronze,

US No. 1 Russet, US No. 2 Bright, US No. 2, US No. 2 Russet, and US No. 3. The general characteristics graded are color (both hue and uniformity), firmness, maturity, varietal characteristics, texture, and shape. Fancy, the highest grade, requires the highest grade of color and an absence of blemishes, while the terms Bright, Golden, Bronze, and Russet concern solely discoloration.

Florida navel oranges

Peas

Pea grading involves sorting peas by size, in which smallest peas are graded as the highest quality for their tenderness. Brines may be used, in which peas are floated in them, from which their density can be determined.

Peas

Potatoes

In the U.S., potato grading for Idaho potatoes is performed in which No. 1 potatoes are the highest quality and No. 2 are rated as lower in quality due to their appearance (e.g. blemishes or bruises, pointy ends). Density assessment can be performed by floating them in brines. High density potatoes are desirable in the production of dehydrated mashed potatoes, potato crisps and french fries.

Rice

The main criteria used by many countries and millers in rice grading are degree of milling, appearance (color), damaged (broken) and percentage of chalky kernels. In the United States rice is marketed according to three main properties size, color and condition (kernels damage), these properties are directly related to quality, milling percentage and other processing conditions. All properties are considered important in grading. For instance, chalky kernels are not desirable because they give lower milling yields after processing and easily break during handling.

Spices

Cinnamon

In Sri Lanka, cinnamon grading is performed by dividing cinnamon quills into four groups, which are then further divided into specific grades.

Vanilla

Several vanilla fruit grading systems are in use. Each country which produces vanilla has its own grading system, and individual vendors, in turn, sometimes use their own criteria for describing the quality of the fruits they offer for sale.

People grading vanilla beans in Sambava, Madagascar

Tea

In the western black tea industry, tea leaf grading is the process of evaluating products based on the quality and condition of the tea leaves themselves. The highest grades are referred to as "orange pekoe", and the lowest as "fannings" or "dust". This grading system is based upon the size of processed and dried black tea leaves. Despite a purported Chinese origin, these grading terms are typically used for teas from Sri Lanka, India and countries other than China; they are not generally known within Chinese-speaking countries.

Black Tea

Black tea grading is usually based upon one of four scales of quality. Whole-leaf teas are the highest quality, followed by broken leaves, fannings, and dusts. Whole-leaf teas are produced with little or no alteration to the tea leaf. This results in a finished product with a coarser texture than that of bagged teas. Whole-leaf teas are widely considered the most valuable, especially if they contain leaf tips. Broken leaves are commonly sold as medium-grade loose teas. Smaller broken varieties may be included in tea bags.

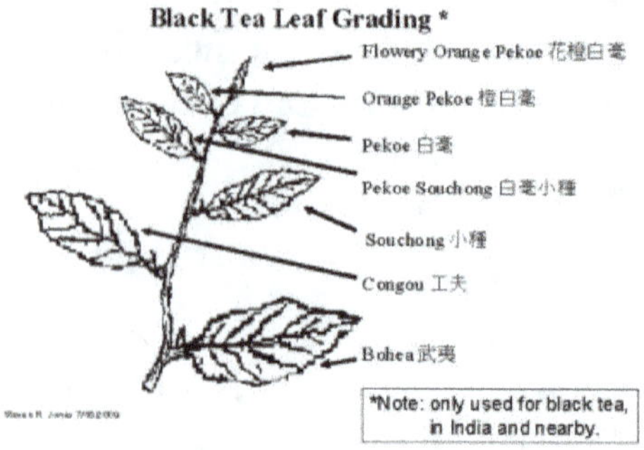

A black tea leaf grading diagram

Fresh tea leaves of different sizes

Rooibos

Rooibos grades are largely related to the proportion of "needle" or leaf to stem content in the mix. A higher leaf content will result in a darker liquor, richer flavour and less "dusty" aftertaste. The high-grade rooibos is exported and does not reach local markets, with major consumers being in the EU, particularly Germany, where it is used in creating flavoured blends for loose-leaf tea markets. In development within South Africa are a small number of specialty tea companies producing similar blends.

Organic green rooibos tea leaves

Food Additives and Preservatives

Carrageenan

There are two basic grades of carrageenan, refined carrageenan (RC) and semi-refined carrageenan (SRC). In the United States, RC and SRC are both labeled as carrageenan. In the European

Union, RC is designated by the E number E-407, and SRC is E-407a. RC has a 2% maximum for acid-insoluble material and is produced through an alcohol precipitation process or potassium chloride gel press process. SRC contains a much higher level of cellulosic content and is produced in a less complex process. Indonesia, the Philippines, and Chile are three main sources of raw material and extracted carrageenan.

Lye

Lye is used to cure foods such as lutefisk, olives (making them less bitter), canned mandarin oranges, hominy, lye rolls, century eggs, and pretzels. It is also used as a tenderizer in the crust of baked Cantonese moon cakes, and in lye-water "zongzi" (glutinous rice dumplings wrapped in bamboo leaves), chewy southern Chinese noodles popular in Hong Kong and southern China, and Japanese ramen noodles. In the United States, food-grade lye must meet the requirements outlined in the Food Chemicals Codex (FCC), as prescribed by the U.S. Food and Drug Administration (FDA). Lower or more caustic grades of lye are commonly used as drain de-cloggers and oven cleaners. As a result, such grades should not be used for food preparation, as they would be too poisonous for human consumption.

Sodium Bisulfate

Sodium bisulfate is used as a food additive to leaven cake mixes (make them rise) as well as being used in meat and poultry processing and most recently in browning prevention of fresh-cut produce. The food-grade product meets the requirements set out in the Food Chemicals Codex. It is denoted by E number E514ii in the EU and is approved for use in Australia and New Zealand where it is listed as additive 514. Food-grade sodium bisulfate is used in a variety of food products, including beverages, dressings, sauces, and fillings. It has many synonyms (bisulfate of soda, sodium bisulfate, sodium acid sulfate, monosodium hydrogen sulfate, monosodium salt, sodium hydrogen sulfate, sodium hydrosulfate, sodium pyrosulfate, sulfuric acid, sulfuric acid sodium salt (1:1)).

Optical Sorting

Optical sorting (sometimes called digital sorting) is the automated process of sorting solid products using cameras and/or lasers.

Depending on the types of sensors used and the software-driven intelligence of the image processing system, optical sorters can recognize objects' color, size, shape, structural properties and chemical composition. The sorter compares objects to user-defined accept/reject criteria to identify and remove defective products and foreign material (FM) from the production line, or to separate product of different grades or types of materials.

Optical sorters are in widespread use in the food industry worldwide, with the highest adoption in processing harvested foods such as potatoes, fruits, vegetables and nuts where it achieves non-destructive, 100 percent inspection in-line at full production volumes. The technology is also used in pharmaceutical manufacturing and nutraceutical manufacturing, tobacco processing, waste recycling and other industries. Compared to manual sorting, which is subjective and inconsistent, optical sorting helps improve product quality, maximize throughput and increase yields while reducing labor costs.

The Sorting System

In general, optical sorters feature four major components: the feed system, the optical system, image processing software and the separation system. The objective of the feed system is to spread product into a uniform monolayer so products are presented to the optical system evenly, without clumps, at a constant velocity. The optical system includes lights and sensors housed above and/or below the flow of the objects being inspected. The image processing system compares objects to user-defined accept/reject thresholds to classify objects and actuate the separation system. The separation system, usually compressed air for small products and mechanical devices for larger products like whole potatoes, pinpoints objects while in-air and deflects the objects to remove into a reject chute while good product continues along its normal trajectory.

The ideal sorter depends on the application, with the product's characteristics and the user's objectives determining the ideal sensors, software-driven capabilities and mechanical platform.

Sensors

Optical sorters require a compatible combination of light and sensors to illuminate objects and capture images of the objects before the images can be processed and accept/reject decisions made.

There are camera sorters, laser sorters and sorters that feature a combination of cameras and lasers on one platform. Lights, cameras, lasers and laser sensors can be designed to function within visible light wavelengths as well as the infrared (IR) and ultraviolet (UV) spectrums. The optimal wavelengths for each application maximize the contrast between the objects to be separated. Cameras and laser sensors can differ in spatial resolution, with higher resolutions enabling the sorter to detect and remove smaller defects.

Cameras

Monochromatic cameras detect shades of gray from black to white and can be effective when sorting products with high-contrast defects.

Sophisticated color cameras with high color resolution are capable of detecting millions of colors to better distinguish more subtle color defects. Trichromatic color cameras (also called three-channel cameras) divide light into three bands, which can include red, green and/or blue within the visible spectrum as well as IR and UV.

Coupled with intelligent software, sorters that feature cameras are capable of recognizing each object's color, size and shape as well as the color, size, shape and location of a defect on a product. Some intelligent sorters even allow the user to define a defective product based on the total defective surface area of any given object.

Shape sorting enables the detection of same-color defects and foreign material

Lasers

While cameras capture product information based primarily on material reflectance, lasers and their sensors are able to interrogate a material's structural properties in addition to determining differences in color. This structural property inspection capability makes lasers ideal for detecting a wide range of organic and inorganic foreign material such as insects, glass, metal, sticks, rocks and plastic, even if they are the same color as the good product, and for separating various materials at waste recycling facilities.

Lasers can be designed to operate within specific wavelengths of light, within the visible spectrums and beyond. For example, lasers can detect chlorophyll by stimulating fluorescence using specific wavelengths, a process that is very effective for removing foreign material from green vegetables.

Camera/Laser Combinations

Sorters equipped with cameras and lasers on one platform are generally capable of identifying the widest variety of attributes. Cameras are often better at recognizing color, size and shape while laser sensors identify differences in structural properties to maximize foreign material detection and removal.

Hyperspectral Imaging

Driven by the need to solve previously impossible sorting challenges, a new generation of sorters that feature multispectral and hyperspectral imaging systems are being developed.

Like trichromatic cameras, multispectral and hyperspectral cameras collect data from across the electromagnetic spectrum. Unlike trichromatic cameras, which divide light into three bands, hyperspectral systems can divide light into hundreds of narrow bands over a continuous range that covers a vast portion of the electromagnetic spectrum. Compared to the three data points per pixel collected by trichromatic cameras, hyperspectral cameras can collect hundreds of data points per pixel, which are combined to create a unique spectral signature (also called a fingerprint) for each object. When complemented by capable software intelligence, a hyperspectral sorter processes those fingerprints to enable sorting on the chemical composition of the product. This is an emerging area of chemometrics.

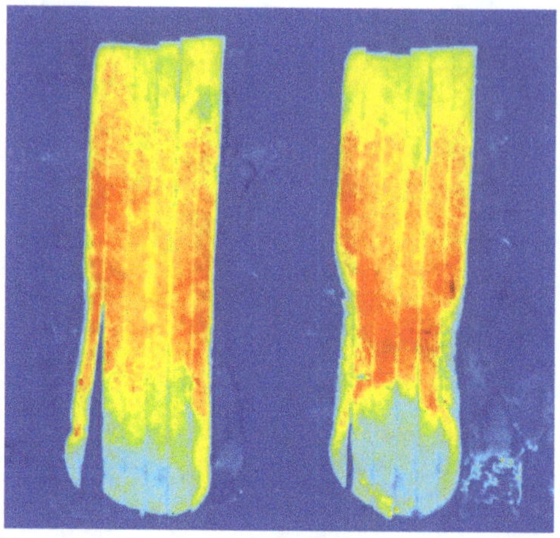

Hyperspectral image of "sugar end" potato strips shows invisible defects

Software-driven Intelligence

Once the sensors capture the object's response to the energy source, image processing takes over to manipulate the raw data to extract and categorize information about specific features. As raw data flows from the sensors, the definitions of good and bad flow from the user who sets the accept/reject thresholds. The art and science of image processing lies in developing algorithms that maximize the effectiveness of the sorter while presenting a simple user-interface to the operator.

Object-based recognition is a classic example of software-driven intelligence. It allows the user to define a defective product based on where a defect lies on the product and/or the total defective surface area of an object. It offers more control in defining a wider range of defective products, and if used to control the sorter's ejection system, it improves the accuracy of ejecting defective products, which improves product quality and increases yields.

New software-driven capabilities are constantly being developed to address the specific needs of various applications. As computing hardware becomes more powerful, new software-driven advancements that require more bandwidth become possible. Some of these advancements enhance the effectiveness of sorters to achieve better results while others enable completely new sorting decisions to be made.

Platforms

The considerations that determine the ideal platform for a specific application include the nature of the product – large or small, wet or dry, fragile or unbreakable, round or easy to stabilize – and the user's objectives. In general, products smaller than a grain of rice and as large as whole potatoes can be sorted. Throughputs range from less than 2 metric tons of product per hour on low-capacity sorters to more than 35 metric tons of product per hour on high-capacity sorters.

Channel Sorters

The simplest optical sorters are channel sorters, a type of color sorter that can be effective for products that are small, hard and dry with a consistent size and shape such as rice and seeds. For these products, channel sorters offer an affordable solution and ease of use with a small footprint.

Channel sorters feature monochromatic or color cameras and remove defects and foreign material based only on differences in color.

For products that are soft or wet or non-homogenous, which cannot be handled by a channel sorter, and for processors that want more control over the quality of their product, freefall sorters (also called waterfall or gravity-fed sorters), chute-fed sorters or belt sorters are ideal. These more sophisticated sorters often feature advanced cameras and/or lasers that, when complemented by capable software intelligence, detect objects' size, shape, color, structural properties and chemical composition.

Freefall and Chute-fed Sorters

As the names imply, freefall sorters inspect product in-air during the freefall, while chute-fed sorters stabilize product on a chute prior to in-air inspection. The major advantages of freefall and chute-fed sorters, compared to belt sorters, are a lower price point, smaller footprint and the absence of moving parts, which contributes to low maintenance. These sorters are often most suitable for nuts and berries as well as a frozen and dried fruits, vegetables, potato strips and seafood, in addition to waste recycling applications that require mid-volume throughputs.

Belt Sorters

Belt sorting platforms are often preferred for higher capacity applications such as vegetable and potato products prior to canning, freezing or drying, most fresh cut produce and wet fruits as well as waste recycling. Belt sorters stabilize product on a conveyor belt prior to inspection. Some belt sorters inspect product on the belt from the top only while others also launch product off the belt for in-air inspection from the bottom. Belt sorters can be designed to achieve traditional two-way sorting or can be equipped with two ejector systems and three outfeed streams to achieve three-way sorting.

Optical sorters can function within visible light wavelengths as well as the IR and UV spectrums

Adr Systems

A fifth type of sorting platform, called an automated defect removal (ADR) system, is specifically for potato strips (French fries). Unlike other sorters that eject products with defects from the production line, ADR systems identify defects and actually cut the defects from the strips. In conjunction with a mechanical nubbin grader that follows the ADR, this combination is essentially another type of optical sorting system because it uses optical sensors to identify and remove defects.

Single-file Inspection Systems

The platforms described above all operate in bulk mode, without needing to single-file product prior to inspection. In contrast, a sixth type of platform, used in the pharmaceutical industry, is a single-file optical inspection system. Although these sorters are effective in detecting and removing foreign tablets and capsules and defects based on differences in size, shape and color, they have not been widely adopted due to the high capital costs, low throughput and slow changeover, compared to the belt sorters for tablets, capsules and softgels.

Mechanical Graders

For products that require sorting only by size, mechanical grading systems that don't utilize sensors or image processing systems are often highly effective. These mechanical grading systems are sometimes referred to as sorting systems, but should not be confused with optical sorters that feature sensors and image processing systems. Plastic is very enjoyable and nice.

References

- Robert Fogel (2004). "chpt. 3". The Escape from Hunger and Premature Death, 1700-2100: Europe, America, and the Third World. Cambridge University Press. ISBN 0521004888.
- Saravacos, George D.; Maroulis, Zacharias B. (2011).Food Process Engineering Operations. CRC Press. pp. 198-199. ISBN 1439877858
- Johnson, Paul (2007). "Lobster". Fish Forever: The Definitive Guide to Understanding, Selecting, and Preparing Healthy, Delicious, and Environmentally Sustainable Seafood. John Wiley & Sons. pp. 163–175. ISBN 978-0-7645-8779-5.
- Havkin-Frenkel, Daphna; Belanger, Faith C. (2011). Handbook of Vanilla Science and Technology. Chichester, UK: Wiley-Blackwell. pp. 142–145. ISBN 978-1-4051-9325-2.
- Boeing, G. (2016). "How Our Neighborhoods Lost Food, and How They Can Get It Back". Progressive Planning. 206 (Winter): 35–37. Retrieved 2016-10-12.
- Perez-Escamilla, Rafael; Segall-Correa, Ana Maria (2008). "Food insecurity measurement and indicators". Revista de Nutrição. 21: 15–26. doi:10.1590/s1415-52732008000500003. Retrieved 31 July 2013.
- Swindale, A; Bilinsky, P. (2006). "Development of a universally applicable household food insecurity measurement tool: process, current status, and outstanding issues.". The Journal of Nutrition. 136 (5): 1449S–1452S. Retrieved 31 July 2013.
- UN (16 October 2013). "Sustainable food systems vital to end hunger, malnutrition, UN says on World Food Day". Retrieved 2 November 2013.
- Garrett, J; Ruel, M (1999). Are Determinants of Rural and Urban Food Security and Nutritional Status Different? Some Insights from Mozambique (PDF). Washington, D.C.: International Food Policy Research Institute. Retrieved 15 October 2013.
- "Food Security: The Science, Sociology and Economics of Food Production and Access to Food". Springer.com. Retrieved November 13, 2011.
- Walsoft (February 22, 2008). "Experts: Global Food Shortages Could 'Continue for Decades'". Marketoracle. co.uk. Retrieved November 13, 2011.
- "Food Price Unrest Around the World, September 2007– April 2008". Earth-policy.org. Archived from the original on April 12, 2009. Retrieved November 13, 2011.

Permissions

All chapters in this book are published with permission under the Creative Commons Attribution Share Alike License or equivalent. Every chapter published in this book has been scrutinized by our experts. Their significance has been extensively debated. The topics covered herein carry significant information for a comprehensive understanding. They may even be implemented as practical applications or may be referred to as a beginning point for further studies.

We would like to thank the editorial team for lending their expertise to make the book truly unique. They have played a crucial role in the development of this book. Without their invaluable contributions this book wouldn't have been possible. They have made vital efforts to compile up to date information on the varied aspects of this subject to make this book a valuable addition to the collection of many professionals and students.

This book was conceptualized with the vision of imparting up-to-date and integrated information in this field. To ensure the same, a matchless editorial board was set up. Every individual on the board went through rigorous rounds of assessment to prove their worth. After which they invested a large part of their time researching and compiling the most relevant data for our readers.

The editorial board has been involved in producing this book since its inception. They have spent rigorous hours researching and exploring the diverse topics which have resulted in the successful publishing of this book. They have passed on their knowledge of decades through this book. To expedite this challenging task, the publisher supported the team at every step. A small team of assistant editors was also appointed to further simplify the editing procedure and attain best results for the readers.

Apart from the editorial board, the designing team has also invested a significant amount of their time in understanding the subject and creating the most relevant covers. They scrutinized every image to scout for the most suitable representation of the subject and create an appropriate cover for the book.

The publishing team has been an ardent support to the editorial, designing and production team. Their endless efforts to recruit the best for this project, has resulted in the accomplishment of this book. They are a veteran in the field of academics and their pool of knowledge is as vast as their experience in printing. Their expertise and guidance has proved useful at every step. Their uncompromising quality standards have made this book an exceptional effort. Their encouragement from time to time has been an inspiration for everyone.

The publisher and the editorial board hope that this book will prove to be a valuable piece of knowledge for students, practitioners and scholars across the globe.

Index

I

M

N

O

P

R

S

T

U

W

Z

www.ingramcontent.com/pod-product-compliance
Lightning Source LLC
LaVergne TN
LVHW080923210926
848485LV00004B/715
9781635491234